BORN ON THE EDGE OF GROUND ZERO
LIVING IN THE SHADOW OF AREA 51

Reme Baca
3110 Grandview St. #1
Gig Harbor WA. 98335

ISBN 978-1-4507-7892-3
LCCN 2011909730

Dedications

This book is dedicated to my son Joel Baca, who spent his vacation in Gig Harbor, using his talent and professional skills in editing and completion of the manuscript and development of the book cover. I always felt that Joel ranked in the top of his industry. Now that I have had the opportunity to observe him in his work, I am convinced that he ranks at the very top. Joel has always been very supportive of our story, and feels that some day when the aliens land their space ship on this planet, their first demand may be that they would like all those craft that our government has confiscated, returned.

Thank you Joel, we are very proud of you.

Acknowledgments

My thanks and appreciation go to Paola Harris, International Journalist, Author, Researcher and promoter of the disclosure project, without whom this book would have never been written, let alone the story told to an international audience. Paola, took the time, at her own expense to travel to Gig Harbor Washington to interview, myself and my wife Virginia, and getting to know us. Her many hours spent interviewing my friend Jose, reviewing and verifying information, and making Col. Corso's notes available to us. Thanks again Paola.

To Ben Moffett, my friend and classmate at the San Antonio elementary school, who made the decision, in the year 2003 to write the story in the Mountain Mail, a Socorro New Mexico newspaper.
Gracias, Hermano Benito.

To Col. Wendelle Stevens, who has passed on, and in his wisdom advised me that the San Antonio crash was a historical event due to the fact that it was the first craft that was recovered intact, included living witnesses, metal and the preservation of the crash site. This type of a combination is extremely difficult to replicate, he said to me.

To Maria Olga Padilla, born June 1940, passed April 2 2011. Both my self, Jeannie, and her survivors will miss her very much. She struggled with Diabetes and kidney disease. Maria never complained, she begin traveling to San Antonio New Mexico, with her husband Jose, helping and supporting the research and development of our project beginning in 2003. Maria came to appreciate life in San Antonio New Mexico, so much that by her wishes Maria was cremated and had requested that she later be buried at the family plot in the San Antonito Catholic Cemetery in San Antonio, N.M.

To William Brophy Jr. who in conversations with his dad, and many hours of research has provided another dimension to our story, and has furnished us with documentation of his dad's service at the Alamogordo Army Air Corps Base, and participation in the recovery effort near San Antonio New Mexico in 1945.

To John Tonan for his support and initial examination and testing of the artifacts. His assistance in the identification of the facilities for the Electronic microscopic examinations and knowledge in working with federal government facilities was invaluable. His experience with metals and as a Science Major the interpretation if the data is most appreciated.

Not everyone who has assisted me in Born on the edge of Ground zero has wished to be identified. Everyone else is exactly as identified in the text.

Preface

The UFO problem has involved military personnel around the world for more than fifty years, and is wrapped in secrecy. Over the years, enough pieces of the puzzle have emerged to give us a sense of what the picture looks like. I have tried to use these pieces to construct a clear historical narrative, focusing on the national security dimensions.

Richard M. Dolan

The 1945 UFO Crash and retrieval of an extraterrestrial object in San Antonio New Mexico is one of the biggest, most intriguing and perhaps the most important story that I have ever covered in my 30 years of investigative journalism. It is clear. It is easy to reconstruct. It is historical in the context of the time. It really happened and once more, it contains witnesses and metals; the verifiable proof that UFO community of researchers is supposedly looking for and which they supposedly call the "smoking gun."

Thanks to the honesty and integrity of Reme Baca now 73 and Jose Padilla now 76, I was able to release the story to the world on December 2, 2010 on the radio show Coast to Coast with George Noory and on several other radio shows. The Padilla/Baca interviews are on the internet and in X-times magazine Italy, and UFO Matrix magazine UK in England but their story has also reached mainland China. It is there for all those who have ears "to hear" and eyes "to see".

One may call this story "the Baca/Padilla gift" to humanity. It is difficult for witnesses of this caliber to completely disrupt their lives, to share such an intimate experience with the world. It affects their families, their work and their privacy. It is a burden of truth that is difficult to live with especially for all these years. Reme

Baca (then age 7) and Jose Padilla (then age 9) truly assisted in an extraordinary event. They not only witnessed a UFO come into Earth's atmosphere, crash, made a half mile trench and stop, but they also interacted with the "creatures" who sent them visual images and showed them the mystery that they were not alone in the universe. Against the precarious backdrop of the atom bomb testing and the presence of the German team of scientists in the middle of the arid New Mexico desert, these two little boys experienced a shift in reality; something that changed their world view forever.

When I began to communicate with Reme Baca in May of 2009, even as a seasoned researcher, it completely changed my world view. Reme Baca is open, forthcoming and his political background and immense knowledge in many fields from metallurgy to sociology and history, gave me a unique perspective to this story. It is important to hear his background, to assimilate the detailed research, the genealogy and the deep analysis of this story, perhaps one of the most important in UFO research. The reader will assist in an event, an event that is part of the greatest cover-up world wide, a cover-up necessary to hide the exotic technologies that have been back-engineered since 1945. Colonel Corso gave his life for this truth. It makes sense that this technological treasure trove is the key to hiding full disclosure. No. Roswell was not the first and only crash. There were many, also in Europe in the 1930's. But this is an opportunity for the reader to read about the 1945 San Antonio New Mexico event which may carry a message for a world racked with anti nuclear messages. These beings may be part of this warning.

Paola Leopizzi Harris
Author: Journalist, Educator And Activist.
Exopolitics: Stargate to a New Reality

It was in this crucible of suspicion and disinterest bred by familiarity that a small contingent of the U.S. Army passed almost unnoticed through San Antonio in mid-to- late August, 1945 on a secret assignment.

Little or nothing has been printed about the mission, shrouded in the "hush-hush atmosphere of the time" But the military detail apparently came from White Sands proving Grounds to the east where the bomb was exploded. It was a recovery operation destined for the mesquite and greasewood desert west of Old US-85, at what is now Milepost 139, the San Antonio exit of interstate 25.

Over the course of several days, soldiers in Army fatigues loaded the shattered remains of a flying apparatus onto a huge flatbed truck and hauled it away. That such an operation took place between about august 20 and August 25, there is no doubt, insist two former San Antonioans, Remigio Baca and Jose Padilla eyewitnesses to the event.

 Padilla, then age 9, and Baca, 7, secretly watched much of the soldiers' recovery work from a "nearby ridge".

Thanks to the work of Ben Moffet, and now thanks to the witnesses themselves, currently in their 70's, the world will know, and see that there was certain extraterrestrial interest in our discovery of the atomic bomb. It opened up a Pandora's box in human history that cannot easily be closed. It put humanity and possible other dimensional visitors in danger of total destruction. San Antonio was more than a crash, a sighting, an event. It was a warning that the military today is not heeding, whether it is in the United States or abroad. If we add the courageous testimony of the Air Force panel led by Robert Hastings and seven retired United States Officers at the Washington Press Club on October 27th of 2010, then we realize the time has come to speak. If these visitors have the capacity of using UFOS to either disable or shoot down our nuclear missiles, then we can see that 70 years later, they are taking action.

Ultimately, 1945 San Antonio case presents a peaceful message, a powerful warning to the planet.

Table of Contents

Chapter Four

Chapter Five

Chapter Six

Chapter Seven

Chapter Eight

Chapter Nine

Introduction

This is a story about two young Hispanic boys of whom you were never meant to know about. A story about two young Hispanic boys, age 7 and 9, growing up at the crossroads of State Highway 85 and U.S. 380 in San Antonio New Mexico. Jose Padilla and Reme Baca literally stumbled across the remains of what they now believe to have been an alien spacecraft, while searching for a cow that was ready to calf on the Padilla Ranch during the war years.

Chapter One

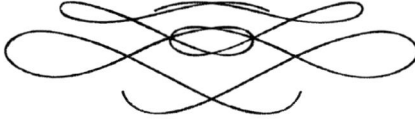

Family History

The San Antonio village has some history related to the development of the Atomic Bomb and nearby Trinity Site, where the bomb was detonated on July 16th of 1945, and the recovery of an Alien Craft that took place involving the army-Air Corps in mid August of 1945, witnessed by Jose Padilla and myself. Among others, the Baca family and Padilla family are descendants of the original Spanish Explorers who established a Government in New Mexico referred to as New Spain.

Cabeza de Vaca was born into the Spanish nobility in 1492. Cabeza de Vaca means literally, "head of a cow". This surname was granted to his mother's family in the 13th century, when his ancestor Martin Alhaja aided a Christian army attacking Moors by leaving a cow's head to point out a secret mountain pass for their use. In the prologue to "La Relacion", his account of his shipwreck and travels in North America, Cabeza de Vaca refers to his forefather's service to the King, and regrets that his own deeds could not be as great. In early 1527, Cabeza de Vaca departed from Spain as part of a royal expedition intended to occupy the mainland of North America. As treasurer, he was one of the chief officers of the Narvaez expeditions.

Within several months of their landing near Tampa Bay, Florida on Aril 15, 1528, he and three other men were the only survivors of the expedition party of 300 men. After finally reaching the colonized lands of New Spain where he encountered fellow Spaniards near modern-day Culiacan, Cabeza de Vaca went on to Mexico City. From there he sailed back to Europe in 1537. After his return to Spain, Cabeza de Vaca wrote about his experiences in a report for King Carlos 1 of Spain. It was published in 1542, under the title of "La Relacion" (The Report).

In March 1599 Don Juan de Onate wrote the Viceroy of Nueva Espana and requested additional soldiers and families to help strengthen the colony that had been established. With approval, the recruitment began in late summer of 1599. Sixty-five Spaniards and twenty-five servants were recruited at Mexico City. By October 1599, recruits, including women, children and servants, were at the outpost of Santa Barbara in the Valle de San Bartolome in Nueva Vizcaya. After some delay, these settlers began their journey to New Mexico in late September 1600 on the route of El Camino Real de Tierra Adentro, the route that Onate had taken in the original expedition of 1598 and which by now had stopping places about every ten miles.

The group arrived at Onate's Cololny on December 24, 1600. Many of these settlers became discouraged and disappointed and left in late October 1601. The families that remained from the second wave of colonization are common ancestors for people with Hispano roots in Colonial New Mexico. Baca: Captain Christobal Baca, native of Mexico City Nueva Espana, and his wife, Dona Ana Ortiz y Pacheco, also native of Mexico City, Nueva Espana, brought three daughters and a son with them to New Mexico. Another son, Alonso Baca was born in New Mexico. Through him, the Baca family name was passed on to the following generations. Padilla: Villasenor (1688; Queretaro, Nueva Espana.)

Jornada Del Muerto

San Antonio is on the main highway between Albuquerque and El Paso Texas. It is eleven miles south of Socorro, the County Seat. "The Journey of the dead, is also known where the route of the dead man begins". The Northeast portion of the basin begins south of San Antonio and adjacent to the White Sands Proving Grounds. It was 90 miles of hell, the roughest and deadliest part of the Camino Real, from Mexico City to Santa Fe, it was the stretch between Las Cruses and San Antonio, a broad, flat valley lacking water, grazing or firewood and it offered no amenities to travelers for over 90 miles

Citizens Of Distinction

Most vacationers are unaware that San Antonio is one of the most historic communities in New Mexico. San Antonio has given many distinguished citizens to our state and our nation, one of the most notable being, Conrad Hilton, son of Augustus Hilton, who became the merchant King of San Antonio.

Conrad was born in San Antonio on December 25, 1887, and died January 3, 1979. San Antonio is where the first Hilton Hotel and other businesses were established. It was the Railroad, and its need for coal that helped develop San Antonio to expand its economic growth. It was during these days that San Antonio enjoyed its widest popularity and growth. As long as the Coal mines at Cartage were working, and the Coke ovens in San Antonio were in operation, the New Town appeared to grow in wealth and Prosperity.

"Augustus Holiver Hilton, Gus" as his wife Mary Lauferseiler called him, had built some rooms behind the adobe store to the West of the railroad station. These were their living quarters. Gus would be gone for days at a time, trading food, tobacco, and whisky for wool, livestock, furs

3

and anything that had trade value. Prospectors looked to him for grubstakes. His Business expanded. He started a stage line between San Antonio and White Oaks. He bought out the partners in the coal mine at Cartage. His friends often referred as the Coal King.

San Antonio took the appearance of the fastest growing town, as storefronts begin filling the empty space on each side of his now large store. The second room that was added to his house was for his son Conrad, who was born on December 25, 1887. An old store building was used as a School. There were 40 School children in school when Conrad started his Primary Grade, and it was bilingual.

Business practically came to a halt with the panic of 1907, and mining came to a halt. Gus decided to change some of the room in San Antonio into Hotel accommodations for salesmen, and this is how the Hilton's started their Hotel Empire. After serving in the State Legislature, Conrad returned to San Antonio with the intentions of opening a Bank. A Charter was granted on September 3, 1913 for the Bank of San Antonio. After World 1 Broke out, Conrad sold the Bank and enlisted for overseas service.

Conrad Hilton was the most distinguished citizen this little town of San Antonio gave to the World. A number of men from San Antonio volunteered for Service in Roosevelt's Rough Riders.

Elfego Baca was one of New Mexico's more colorful characters, regarded as both a bad guy and a good guy, Baca's life was full of controversy, drama, and humor. He pursued careers as Lawyer, private detective, real estate sales, newspaper publisher, sheriff, county clerk, school superintendent, mayor, district attorney, and candidate for governor of New Mexico and for U.S. House of representatives.

4

Baca was also known as a pioneer gunfighter. Baca was born in Socorro, New Mexico, although his father (Francisco Baca) moved the family to Kansas for part of his youth, Elfego returned to Socorro and from there launched into his political career. He was elected Socorro County clerk in 1893, and served in this capacity until 1896. During his tenure as county clerk, Baca was admitted to the bar. From 1896-1898, he was mayor of the city of Socorro. He then became school superintendent for Socorro county from 1900-1901.

Governor Otero appointed him district attorney for Socorro and Sierra counties in 1905. The last elected office that he held was sheriff of Socorro County, 1919-1920. It is understood that Elfego and Billy the Kid were friends, and it was Elfego that took Billy to a Mexican dance in Socorro, where Billy met a lady that was to be his future spouse. Baca practiced law in Albuquerque during his later years. He died at his home in Albuquerque on August 27, 1945.

The Village

In 1945, San Antonio was not a middle class town, for it was much closer to the third world. It borders San Antonito as a small village to the South, and Laborcita, a hamlet to the north of San Antonio. We grew up in an exciting time. We grew up in this little village where everything is scattered around and there's a wrecked car without an engine or wheels rusting in someone's back yard, a forgotten antique gas pump stands on the side of a street without a gas station around it. There's a long and flat abandoned building with broken windows next to a restaurant. It's impossible to figure out what it ever housed.

My dad was a "sharecropper" before he worked at the Veterans Hospital as an attendant. He was responsible for the crops from tilling to harvest, and he got the smallest share. We lived in a farmhouse, complete with windmill, hand dug well and an irrigation ditch close by. In 1945, the town was kind of primitive, compared to present times. San Antonio was established sometime in the 1660's and was named after the Piro Village and mission of San Antonio De Senecu. A post office was established in 1870.

The inhabitants of San Antonio as described in the 1885 Census, included farmers, stock raisers, saloon keepers, shepherds, lawyers, a silversmith, cobbler, blacksmiths, boarding house owners, railroad station agents, telegraph operators, restaurant owners and hotel clerks. Before the Civil War, coal was discovered ten miles to the east of the village and it was mined to supply the needs of nearby Fort Craig and the surrounding communities. The flood of 1929 partially destroyed San Antonio and the depression almost completed the task.

Out Door Plumbing

When Jose and I were young, we had no electricity, running water, indoor plumbing, or health facilities and public services that other small towns had. The cook stove was fired with wood. In cold winters, I would walk along the Santa Fe Railroad tracks, picking up coal that fell out of the railroad cars and put it in a bucket. This is what was used to stoke up our coal stove to keep our family warm in winter.

In addition to that, I remember that the only telephone system that existed for us, consisted of two empty milk cans, with the top removed, strung end to end with a piece of twine. The sender would speak into the milk can at one end of the twine and the receiver would listen at the other end.

Saturday night was bath night, and water would be hauled from the well, heated on the wood stove and poured into a number two tub, where you would take your bath.

Lighting was provided by kerosene lamps. Food was what you grew in your garden. The cement tank stored water pumped by the windmill and it was used to water the vegetable garden and by family and school friends on hot summer days to cool off, when temperatures would get up to 110 degrees in the shade. Under the windmill was a hand dug well, about eight feet deep, complete with salt cedar casing to keep the sides from caving in, where a well point had been sunk down about fifty feet below the ground. In this well is where the pumping mechanism for the windmill was located to keep it from freezing during the cold winter months. The windmill would turn this mechanism and move a rod connected to a valve inside the cylinder up and down creating a vacuum inside the cylinder, which would suck water up and pump it into the cement storage tank.

The bathroom was an outhouse that was located in our back yard, a genuine WPA (work projects administration) built one. The WPA built several models, ours was a single seat. They didn't come with potty wipe. In those days, we had to be innovative. We used the two major catalogues for toilet paper. Come to think of it, this may have been one of the most effective methods of dealing with unwanted advertising. These were the dream books of the era.

Across the dirt road was an old adobe building with two rooms that belonged to our family. One room was used for grain storage and the other one for hooks for saddles and harnesses for horses and a plow. The grain storage room was usually available for a sheepherder friend of my dad's, who used it in the fall. He was part of the crew that brought the sheep to the stockyards for shipping to the slaughter houses. Out side the building was a horse drawn wagon, a corral large enough for two or three horses, a pile of hay, scratchy and Smelly, but wonderful to tumble in.

The kitchen at our house was dark and warm, the curtains were drawn against the dark outside, against the rain driving past the house from the northwest. An arrangement of dry holly-hock flowers set on the table, yellow pale gold blossoms seemed to add light to the room. A detectable smell of bread crumb pudding was coming from the oven. Our Thanksgiving would consist of refried beans, tortas, fried potatoes, white wheat flower tortillas and bread pudding. The only turkey in our house was the picture that had been copied on the Hectograph and colored by me at School.

Reunited

I moved to Tacoma Washington in 1955, while still in high school, Jose moved to Rowland Heights, California about a year before. A few years ago, after more than four decades apart, we were brought together thru the internet. I was doing research on our family genealogy and made contact with someone named Padilla, and it turned out to be Jose Padilla's son. It was then that our interest in the most intriguing event of our childhood was rekindled. I contacted Jose and we began trying to catch up on the events that had taken place since we had left San Antonio.

Classmate

Another event included a former classmate Ben Moffett who influenced this story. At the time he was a Journalist for the Mountain Mail, covering Socorro and Catron counties in New Mexico. Jose had sent me a newspaper clipping that included an article written by Ben Moffett. Apparently Ben Moffett while visiting the San Antonio Elementary School, had mentioned his other classmates, but had not mentioned us. After a conversation with Jose, I decided to give Ben a call. Ben answered the phone, and after a short conversation, I asked him about the omission. Ben's response was, "I really meant to include you," but I had forgotten your names, it has been so long. After various conversations on the phone, and email, catching up on events covering several years since they had talked, Ben kept asking me what I was doing now that I had retired. I tried my best to evade him, being aware that one of most effective methods of losing a friend is to mention anything that could be interpreted as UFO's. Finally, Ben continued to press me, and I decided to tell him about my research regarding a crash of an unknown object on the Padilla ranch in 1945. Ben's immediate response at the time was, "I don't believe in UFO's." We continued our contact by phone and email, talking about old times while attending San Antonio elementary school, where we played on the same basketball team, and the great coach, Mr. Pete Eaton was and the Trophies we won for him. Unexpectedly, I received a call from Ben, stating that after some research, and checking with his editor, a decision had been made that he should print the story in the Mountain Mail, a local Socorro County Newspaper, where he was employed, and so the story began.

The Manhattan Project

Just before dawn on July 16, 1945, scientists detonated the world's first Atomic Bomb at the Trinity site, some 20 miles southeast of San Antonio, New Mexico. Security precautions were rigid. In March, Dana P. Mitchell, Assistant Director at the Los Alamos Laboratory, issued terse, precise travel instructions: "The following directions are strictly confidential and this copy is to be read by no one but yourself." "You are to turn this copy in to me personally on your return to the site," the memo read, and continued with specific directions and mileages for reaching the site. "Under no condition," it went on," "when you are south of Albuquerque are you to disclose that you are in any way connected with Santa Fe. If you are stopped for any reason and you have to give out information, state that you are employed by the Engineers in Albuquerque, under no circumstances are telephone calls or stops for gasoline to be made between Albuquerque and your destination." Travelers were then instructed to "stop for meals" at Roy's in Belen, which is on the left-hand side of the main road going south. If you leave the site at 7 a.m., you should make this stop around lunch-time.

Plenty of local procurement problems remained. First there was communications. Only five people on the project were allowed to telephone between Trinity and Los Alamos and these calls were routed to Denver, on to Albuquerque and finally to San Antonio, New Mexico. Teletype service was so bad, Van Gemert recalls, that you never knew if the test site was asking for a tube or a lube job. It soon became evident that the best way to communicate was to send notes back and forth by the truck drivers. At least two and very often as many as ten trucks left Los Alamos every evening after dark to avoid both the blistering desert heat and unnecessary notice, and arrived at the test site early the next morning. Almost always there was a stop to be made at the U.S. Engineers yard in Albuquerque to pick up items addressed to Prof. W. E. Burke of the University of New Mexico's physics department, who served as a blind to avoid a connection between the items and Los Alamos. "We'd get things to Trinity any way we could," Van Gemert says. Some of the ways were devious.

A carload of telephone poles was desperately needed at the test site and no freight train was traveling fast enough to get it there in time. After considerable urging the Santa Fe railroad consented to attach the car to the rear of the Super Chief and sped the cargo to Albuquerque. Another time, for lack of freight space, 24 rolls of recording paper were luxuriously ensconced in a Super Chief drawing room for the trip from Chicago.

To supplement the special items, the procurement people established a complete technical stockroom at the test site early in the game and trucked the entire stock from Los Alamos. The stockroom, known officially as FUBAR (Fouled Up Beyond All Recognition), was manned by enlisted men who used their spare time to manufacture the face shields needed to protect observers from the test blast. The shields were made of aluminum sheets, mounted on a stick handle, with welders' goggles for a window.

There never seemed to be enough people to take care of all the work to be done on the test preparations and those who were available, from mess attendants to group leaders, worked at a fever pitch. A ten hour day was considered normal and it often stretched to at least 18 hours. In the spring of 1945 a big part of the laboratory was reorganized to take care of the test and many people found themselves involved in activities far removed from their normal duties. John Williams, the High Energy Physicist, took the responsibility for construction and servicing of the base camp. John Manley was wrapped up in neutron measurements as a Research Division group leader when he suddenly found himself in charge of blast measurements for the test. "I didn't know anything about blast measurements," he recalled 20 years later. "We'd never done anything like that before." But talent is talent wherever it is found and the displaced crews managed expertly and efficiently to bring their remarkable tasks to a successful conclusion under extreme pressure.

Throughout the spring and summer there was a constant stream of personnel traveling between Site Y and Trinity in a motley assortment of busses and cars, some of them barely able to make the long, monotonous trip. Even so, by mid afternoon when the travelers reached the little junction town of San Antonio, most of them were hot, tired and thirsty and Jose Estanislado Miera's bar and Service station became a popular, if illegal, stop.

The Owl Bar and Café in San Antonio was a watering hole for the scientists, engineers, and soldiers who made the trip from Los Alamos, some 140 miles away, traveling to and from the test site. The owner, Estanislado Miera provided food and shelter including the use of his telephone to them on their long trips. Some cabins (sleeping facilities) were reserved for their long overnight stay. The cabins windows would be covered with newspaper and cardboard from the inside.

Miera still remembers the unusually heavy traffic in those days. One of his customers, John Manley, remembers "Miera's wall of a bottle collection." "He had the whole south wall of his place lined with liquor bottles," Manley reports. "We used to worry an awful lot about that, if our big blast traveled that far, that's the wall it would hit." Luckily it didn't. Their very presence in San Antonio over several months should have been a red flag that some secret government project was happening nearby. Some of the people in San Antonio were aware that something was going on and when they got up early that July 16th Monday morning for a sight that they would never forget. A Soldier that was assigned to an observation point at Estanislado Miera's Café informed him on the morning of the test, that if he stepped outside, "You'll see something the world has never seen before."

Estanislado had been advised by one of the Los Alamos engineers having lunch there to take down his bottle collection that was sitting against the wall and store it for a few days, and was he glad he did. A vehicle had been provided for State Policeman Apodaca by the military, in the event that they needed to evacuate the population of

San Antonio should the radioactive cloud change directions and pass over the town.

Jose Padilla and his mother experienced the first atomic bomb coming to life at the trinity site some eighteen miles to the Southeast of their ranch house too. Jose was nine years old at the time and his mother, Inez had just seen her husband Faustino off to work. Jose was drinking a cup of hot chocolate when the flash, the sustained deadly sound and the heat wave came. His mother peeked through the door at the flash of light. As a result, she sustained permanent loss of sight in that eye.

I was home asleep when I felt my bed shake, and a bright light came into my room. My bed felt like it was jumping across the room. This is what woke me up. I had been shaken out of my bed. It seemed like a train was coming thru the front door of the house. For a few seconds, I though it was day light, time to get up. That's what I was thinking as I picked my self off the floor. This also woke my mother Eva up. She assured me that everything was all right to go back to sleep and that it must have been the result of the thunderstorm that we had experienced during that night. A few weeks before this incident, in the late morning, we had heard a loud explosion, and seen a plume, but we were at a loss for an explanation.

Ben Moffett, retired chief of public affairs for the Rocky Mountain Region of the National Park Service, was a boy living in San Antonio at the time. "I may be the only person who slept through the blast, but I remember it vividly because my parents (John B. and Regina Moffett) ran from the kitchen where I was asleep and woke me up to see if I was safe," Moffett said. "I was a month short of six years old, but I remember the panic in their eyes, something I had never seen before."

Later in the day we went to Socorro to peddle vegetables house to house and it was all anyone could talk about. In 2005 a group of San Antonio natives gathered for a discussion on their experiences at New Mexico Tech's Skeen Library. The talk was organized by San Antonio native Ricardo Padilla Reyas. Lucille Catherine Miera told the audience she remembered military men in San Antonio. "There were other men, too," she said. "I met many of them, and was told later that Oppenheimer was there. I may have met him, but I was only 13 at the time, so I'm not sure."

She said the men were gone during the day, and came back at night. "No one said where they worked other than they were just working on a project," she said. Miera remembered overhearing some of the GI's talking on the telephone. "They said they were talking to their girlfriends. Oh, I've got to call Mary Ann," they would say. "But the things they were saying didn't sound like it. They were saying things like how many of this or that, and what are the times? Something you wouldn't expect someone to be talking to their girl friends about. They would say I've got to call my other girlfriend, so-and-so," Miera recalled. "We figured they were using some kind of code. Or they had lots of girlfriends. " She said her grandfather had rented out cabins for the service men and government people, and there were also trailers and army tents in the vicinity.

"Friday night was always movie night at the recreation area, and all the soldiers were there like usual, eating popcorn," she said. "Saturday was normal, but Sunday was quiet." "We had a flashlight, and my grandmother would sometimes wake up by shining it in my face," she said. "That morning I woke up because of a bright light, and I said, "Turn it off, Grandma," but she wasn't there and there was no flashlight. The whole room was lit up." "I didn't know what was happening." she said. Her grandmother came into the room and told her to stay inside. "But I went outside to see anyway," she said. "Everything was real bright, like a halogen lamp. The little trailers and tents were all gone."

Juanna Gonzales Odeb remembered waking up at 5:30 that morning. There were no lights, no phone, no nothing, she said. It was already getting to be daylight. You could see it. Abig mushroom cloud. "I remember thinking we were being bombed by the Japanese," she said. Cecelia Padilla Woodward said she was playing in her yard with her sisters when the bomb exploded. "I remember we were playing outside," she said. She remembers the farm animals being sick afterward." There was a kind of mucus coming out of their beaks," she said. "Some of the cows lost their hair."

Ana Lee Padilla Montoya said her husband Atreraclio Montoya worked at the trinity site on the tower that held the bomb. "They never told him what it was for," she said. About 200 local workmen helped construct the tower, a fact which hasn't been made known to the general public. "But, all you see in the old photographs are military workmen," he said. "They kept the locals out of sight when pictures were being taken." Ana Lee said after the test her husband went to work for the Bureau of Land Reclamation. "He was never told to keep secret about his job," she said. "He just started looking for a new job."

What she remembers that morning was seeing the cloud. "We saw the mushroom cloud and all that red light," she said. After the blast she said some of herb sisters had cysts and other health problems, and she had three miscarriages. "We didn't find out the radiation dangers until after the war," she said. "They should have notified us about the risks."

Declassified documents have revealed that the military knew about many of the dangers of fallout to the local population. "The military was apparently afraid that any request for residents to move would give away the secret nature of the project," Reyas said. "Although they knew about the risks of radiation, they felt it wouldn't be high." He said subsequent documents state that future tests should be held 150 miles from any populated areas.

"That's why they moved above ground testing to Nevada," he said. "This indicates they knew the radiation was more dangerous." Reyas said the first choice for testing the WMD was California. "But since General George Patton did his tank training there, Oppenheimer would not consider it," he said. "He hated Patton personally, and didn't want to have to deal with him." Visitors are allowed to enter and exit White Sands Missile Range's Stallion Gate off Highway 380 unescorted during the open house, although identification is required to enter. During the free open house visitors can personally inspect ground zero where the July 16, 1945 20 Kiloton explosion occurred, and take a shuttle bus to the MacDonald ranch compound where the bomb was assembled. The Trinity Site is open to the public only two days each year; on the first Saturday of April and October. The Stallion Gate turnoff is 12 miles west of San Antonio.

Benito said, "... good job, John Larson." Socorro and environs have more history than you can find in one undersized area. From Socorro to the Texas State line over Highway 380, you have Robert Oppenheimer and his gang of Physicists, Billy the Kid, Lew Wallace, Peter Hurd, Kit Carson, Elfego Baca, Robert Goddard, Robert O. Anderson, Black Jack Pershing and Smokey Bear, just to name a few. I left out Conrad Hilton. "There are more," Benito said.

Chapter Two

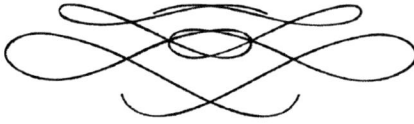

A Profound Influence by Melchora

Folks who live well beyond the lifetime of their colleagues often are not much remembered when they pass on. Not so educator Melchora Gonzales, who died at her home in Socorro on April 19 at age 99. Ms. Gonzalez left behind hundreds of former students and teachers she inspired, each of whom carry a treasure trove of memories about her and remember her grassroots efforts to educate youngsters in Socorro county from September 1929 until her retirement in may 1970.

Although her name may be permanently on file in the public record for her highest ranking job, Socorro County School Superintendent, she will best be known as a grade school teacher who inspired impoverished rural students, including this writer, to excel and pursue a better life.

"She started teaching in San Marcial in 1929, the year the river flooded the town," said her sister, Placida Gonzalez of Socorro. "From there she went to Tokay, San Pedro, La Joya, Escondida, and San Antonio and then she ran for superintendent of schools."

"Somewhere in there she also taught at dusty," according to Yolanda Bianchi-Cook, a cousin and first grade student of "Miss Gonzalez", a title of respect Yolanda never quit using despite their kinship. Dusty was a ranching and mining community in the southwest corner of Socorro County.

Why Gonzalez moved from school to school during her early career is not entirely clear, but it probably was the result of outlying communities becoming ghost towns or consolidating with larger schools. San Marcial's school never reopened after the flood and Tokay also disappeared. San Antonito's school burned down and the students were moved to San Antonio. San Pedro's students eventually were bused to San Antonio.

In 1942, at about age 37, Republican Gonzalez defeated Martin Lopez Jr. by the count of 2117-1757 for the job of county school superintendent, according to State Library archives. It would be interesting to look at the Precinct results, but it's probably safe to assume that large majorities for Gonzalez came from the rural school precincts where she worked.

In 1944 Gonzalez, born in San Antonito, returned "home" to San Antonio Grade School. Yolanda Bianchi and I were two of her first-grade students. I didn't know until I talked to Yolanda this week that she was then only four years old, head-started by Gonzalez. Because of her early start, Yolanda graduated from Socorro High in 1957 at the tender age of 16 years, one month.

I was also head-started by Gonzalez in a different way. A farm kid with little access to books, School Superintendent Gonzalez loaned my father and mother three textbooks to study over the summer "Look and See". "Come and Go", and "Work and Play", about a middle class, five-in-the family "Dick and Jane" series produced by Scott Foresman & Co. in 1940.

Gonzalez must have known she was headed for San Antonio as my teacher when she loaned the books out to my parents. I mastered them over the summer and in the classroom and developed into a confident reader, a fact that

I am sure enabled me to become the first college graduate in my family. I am also sure that it gave me an understanding of the importance of reading readiness, a value shared by my wife, Lesta. Through her nurturing, Melchora Gonzalez was a significant contributor to our own children's success. Two are educators, Pam, is a high school English teacher in Alamogordo, and Marc, a professor of philosophy at the University of Wyoming.

I am not alone in my assessment of Gonzalez. "I remember she gathered us in a semi-circle in tiny little chairs," said Frank Chavez, a Las Cruces attorney and another 1944 classmate. "There was a large (Dick and Jane) book in the middle and she used a pointer to help us. She gave us the fundamentals, the basics to build an education. She expanded our horizons."

George Aguilar, another 1944 classmate who lives in Socorro, "She was just a good teacher." Aguilar's wife Ida Mae, who didn't train under Gonzalez, says she is nonetheless surrounded by Gonzalez advocates including her husband, her father, Ramon L. Baca, and her daughter Vicky Gonzales.

It wasn't only students who were fans of Miss Gonzalez, but other teachers, too. She recruited Anselmo Valenzuela of Luis Lopez to the teaching ranks. "I was a third-year student at the University of New Mexico and working in a lumber yard when she invited me to become a teacher," he said.

Valenzuela later taught some of Gonzalez's students, including Yolanda and me, at the fifth-grade level at San Antonio, but he recognized he had a brighter future in another field and left after a short while. Valenzuela is now a retired engineer living in California.

Gonzalez not only recruited people like Valenzuela into the profession but got youngsters interested in education as youngsters. Yolanda's sister, Carmela

Bianchi-Walentowski of Socorro and Albuquerque, was a Gonzalez student who became a teacher in Los Lunas, and she credits Miss Gonzalez with moving her in that direction.

Success stories are ubiquitous in that 1944 class which included David Ortiz, who trained at New Mexico Tech and became a high level engineer with the federal government, and Remigio (Reme) Baca, who became an insurance executive and political kingmaker in the State of Washington as a strategist for the successful Gubernatorial campaign of Dr. Dixy Lee Ray in the late mid nineteen seventies, and mayor Doug Sutherland in the 1980's, and many others.

At my mother's insistence, I created a scrapbook of my first grade papers, and it exists today in a trunk somewhere in my home, complete with Gonzalez's markings and comments. Probably there are also report cards issued by Gonzalez, which contain Abraham Lincoln's words: "I will study and prepare myself, and then perhaps my time will come."

Thanks to Miss Gonzalez's efforts, Lincoln's words about preparation happened for so many of us in that class, and in many of her classes before and after.

A good thing too, the summer after that school year, July 16, 1945, the world's first atomic bomb exploded at Trinity Site, some 25 miles from the San Antonio schoolhouse, ushering in the Atomic Age. We needed all that Gonzalez-inspired preparation, because the world's knowledge base has been increasing exponentially every since and the three "R's" have been increasing important.

Thank you, Melchora Gonzalez, (Dec. 19, 1905- April 19, 2005) for a century of service.

End of War

Three weeks later, on August 6th and 9th the United States brought the 2nd world war to a dramatic end by using the bomb to destroy the Japanese cities of Hiroshima and Nagasaki. On august 6th, the world first learned that the trinity event, which had frightened the people of San Antonio, was not "an ammunition magazine containing high explosives and pyrotechnics" as the military had earlier reported in front of the Owl Bar and Café. It was and atomic bomb, "death, the destroyer of worlds", in the words of project Physicist J. Robert Oppenheimer. When the bomb was detonated down at Trinity, residents for hundreds of miles knew something extraordinary had happened. Light was seen in El Paso Texas. A blazing flash was reported in Santa Fe. Windows rattled in their frames as far as Gallup and Silver City, over two hundred miles away. Men who were part of the project, watched from arroyos and slit trenches. Up on Compania Hill just a few miles east of San Antonio, and some twenty miles from ground zero, where some of the Scientists and Engineers had applied sunburn lotion to their faces in the predawn dark, they saw it, they could see it from San Antonio and from up on Chupadero Peak. On shortwave Radio, the voice of America played the "Stars Spangled Banner" just before detonation.

Some would say that what occurred at Trinity was one of the greatest moments in history. The discovery of the wheel, fire, and electricity were paled by comparison. It was the birth of a new religion, one in which the end of the world might be near, but no repentance would provide for salvation. It was Robert Oppenheimer who gave the test its name of Trinity, and Oppenheimer who said later that what came to his mind, as the arroyos and hills were bathed in seething brilliance as the purple radioactive glowing cloud hung there over the desert, there was a phrase from

Hindu scripture, from a scene in the Bhagavad-Gta. Vishnu, hoping to motivate the Prince to behave dutifully, takes on his thousand-armed form and says, "I am become Death, the destroyer of worlds." Many of its makers would later memorably describe what they witnessed as the convection stem of dust chased the final cloud of smoke towards the morning star.

Trinity Test, July 16, 1945-Radiation Monitoring Source: U.S. National Archives, record Group 77, Records of the Office of the Chief of Engineers, Manhattan Engineer District TS Manhattan Project Files, folder 4, "Trinity Test".

The Trinity test, on July 16th, 1945, was a spectacular success. A 6 Kilogram sphere of Plutonium, compressed to super criticality by explosive lenses, exploded over the New Mexico desert with a force equal to approximately 20,000 tons of TNT. This report, by Col. Stafford Warren, Chief of the Manhattan Project's Medical Section, shows that the potential for radioactive fallout from the test was an important concern.

Warren's report shows that fallout from the test exposed a family living 20 miles from Ground Zero to dangerous levels of radiation. By July 27, General Groves' office diary shows, the radiation monitors became so concerned that they asked permission to talk to the family to see how they felt.

TOP SECRET
By Authority of the District Engineer
Per K D Nichols

THIS DOCUMENT CONSISTS OF 3 page (S)
NO> 1 OF 4 COPIES, SERIES A

21 July 1944

To: Major Gen. Groves

SUBJECT; Report on Test II at Trinity, 16 July 1945

1. The test was performed two days ahead of the tentative schedule because everything of importance to the test was ready.

2. A study of the weather indicated that a variety of wind conditions at slow speeds going in general N.W. could be expected with directions and speeds at different levels for 16 and 17 July 1945. These slow winds could be advantageous in localizing the outfall of active material from the cloud to the site and nearby desert areas. They would also dilute the outfall most effectively in the early hours of the life of the cloud when it would help the most. The monitoring problem would be sores, however, because of the wide area covered.

3. In the two days available, the population of the surrounding areas was located by G-2 on large scale maps for a radius of 75 to 100 miles. The deserted areas corresponded fortunately to the most probable courses of the outfall from the cloud as predicted by the directions of the winds at the various altitudes. Troops under Major Palmer were available if monitoring indicated that evacuation was necessary

4. At zero minus five hours, five cars with Dr. J. Hoffman in charge were stationed with Major Palmer and troops at the outlet road near the East-West Highway #380. They were in radio communication with Base Camp and Post #2. Outlying monitor cars were in San Antonio, Roswell, Carrizozo and Fort Summer to cover these areas in case the speed of the cloud was greater than predicted.

5. Dr. Aebersold was in general charge of the monitoring at Base Camp and the three shelters at

10,000 yards, with local telephone and radio communication. There was a technician monitor and doctor in each shelter and at Base Camp.

6. Dr. Hempelmann in charge of all the monitoring programs was at S 10,000, the center of communications and final decisions including Brig. Gen. Farrell, Dr. Oppenheimer, Dr. Bainbridge, Hhubbard.

7. This officer acted as liaison in a secondary communication center in Base Camp. Lt. Friedell was located with G-2 at Albuquerque at another communication center via Long distance for controlling the field monitoring in case Base Camp communications broke down. All groups were keyed in by identical maps showing preliminary locations, monitors, their presumed course, the two possible paths of the cloud, WNW and NNE. Monitors, their presumed course, the two possible paths of the cloud, WNW and NNE (depending upon the altitude which it reached) houses and nearby ranges, etc. Albuquerque as another communications center: Preliminary locations of the monitors, their presumed course, the two possible paths of the cloud, WNW and NNE (depending upon the altitude which it reached) houses and nearby ranges, etc.

8. Accessory equipment and other preparations were in keeping with the preliminary plans submitted in the preliminary report.

9. The shot was fired at 0530 on 16 July 1945. The energy developed in the test was several times greater than that expected by scientific group. The cloud column mass and top reached a phenomenal height, variously estimated as 50,000 to 70,000 feet. It remained towering over the northeast corner of

the site for several hours. This was sufficient time for the majority of the largest particles to fall out. Various levels were seen to move in different directions. In general the lower one-third drifted eastward, the middle portion to the west and northwest, while the upper third moved northeast. Many small sheets of dust moved independently at all levels and large sheets remained practically in situ. By zero plus 2 hours, the main masses were no longer identifiable except for the very high white mass presumably in the stratosphere.

10. By 0800 hours the monitors reported an area of high intensity in a canyon 20 miles northeast of zero. Since this was beyond the tolerance set and equally high intensities were expected in other areas, four more monitor cars were sent into this northeast area from Base Camp. The roving monitors in this area were each accompanied by a trooper in a 4 wheel drive and authorized to evacuate families if necessary. At no house in this whole north and northeast area between 20 miles and 40 miles from zero was a dangerous intensity found. The highest intensities, fortunately, were only found in deserted regions. The highest found is shown in detail attached #1. Intensities in the deserted canyon were high enough to cause serious physiological effects.

11. The distribution over the countryside was spotty and subject to local winds and contour. It skipped the nearby #380 (20 mi. N.E.) except for low intensities which were equaled at twice and three times the distance. It is presumed that the largest outfall occurred in the N.E. quadrant of the site. This can only be explored by horseback at a later date.

12. The monitors all took considerable risks knowingly and many have received exposures of considerable amounts, i.e. 8r total. This is safe within a considerable margin. They should not be exposed to more radiation within the next month.

13. The dust could be measured at low intensities 200 miles north and northeast of the site on the 4th day. (Attached #2) There is still a tremendous quantity of radioactive dust floating in the air.

14. Neither the Base Camp or the shelters were contaminated very much.

15. Partially eviscerated dead wild jack rabbits were found more than 800 yards from zero, presumably killed by the blast. A farm house 3 miles away had doors torn loose and suffered other extensive damage.

16. Details indicating blast, heat and other effects cannot be worked out until the area around the crater "cools down". It is this officer's opinion however, that lethal or severe casualties would occur in exposed personnel up to two miles from a variety or combination of causes, ie., blast, heat, ultraviolet and missiles. The light intensity was sufficient at nine miles to have caused temporary blindness and would be longer lasting at shorter distances. Several observers at 20 miles were bothered by a large blind spot 15 m after the shot. The light together with the heat and ultraviolet would probably cause severe damage to the unprotected eyes at 5-6 miles; damage sufficient to put personnel out of action several days if not permanently. All of the personnel obeyed the safety precautions during the test so that no such injury resulted.

17. A great deal of experience was obtained on the requirements for quick and adequate Monitoring. Excellent radio communications, good transportation and better and more rugged meters are required.
18. It is this officer's opinion based on the damage to "Jumbo" (2400 ft), the extent oft the glazed sand area (up to 500 ft..), the extent of the cleaned off area (about 1 mile), the farm house (at 3 miles) that this explosion was a great many times more violent than the 100 ton test. "Conservative" estimates by the scientific groups put it at least equivalent up to 10,000 tons of T.N.T.
19. While no house area investigated received a dangerous amount, i.e., no more than an accumulated two weeks dose of 60r, the dust outfall from the various portions of the cloud was potentially a very serious hazard over a band almost 30 miles wide, extending almost 90 miles northeast of the site.
20. It is this officer's opinion that this site is too small for a repetition of a similar test of this magnitude except under very special conditions. It is recommended that the site be expanded or a larger one, preferable with a radius of at least 150 miles without population, be obtained if this test is to be repeated.

Colonel Stafford L. Warren
Chief of Medical Section
Manhattan District
SLW/fp
cc/ Maj. Gen Groves (2)
R. Oppenheimer (1)
Col. Warren (1)

27

The two attachments, diagrams of radiation "hot spots" 20 miles from ground zero, are not reproduced here. The diagrams show radiation levels on U.S. Highway #380, 20 miles NE of ground zero, as "Intensity approximately 50 r total for 10 miles along this highway".

The second attachment is titled "Location of Hot Canyon 2 miles from zero along N.E. path of cloud" The canyon was 5 miles east of Bingham. This is the description of radiation levels:

"Hot canyon" 1.1 miles east of road junction.		
15.	Or/hr at 0300 hours after zero)	Total dose = 212 r
14.	Or/hr at 0330 hour after zero)	
6.	Or/hr at 0830 hours)	To 230 r
	0.6r.hr at 3600 hours (after rain)	
One mile east of the canyon, the diagram notes the location of a house with "family with 1 child", with radiation dose described as follows: House (with family) 0.9 miles beyond "Hot Canyon"		
0.4 r/hr at 3600 hours after zero & after a rain	Accumulated total dose 57-60 r	

4:15 PM. General Groves called Col. Warren re Friedell making measurements on certain family out there. Gen. Groves suggested that Col Warren call Davies in New York and get the matter straightened out because there seems to be a difference in stories between Daley, Hemplemann and Davies. Gen Groves also asked Col Warren to tell Davies that he was released as far as the General was concerned. Lt. Davies from MSA, N.Y. called Gen Groves and advised Gen Groves that Lt. Daley called him and informed him that Friedell's boys had made some further observations and are concerned about one family to the extent that they want to get in touch with tat family to see how they feel. They called me to ask about the legal end—told them there was nothing I could do about it. Told them to confer with Col. Warren and you. I am sure nothing has been done yet. The call came in a few minutes ago and thought I would relay the message to you. Davies advised the General that he would be at Maj. Tandy's office at MBA and will stay there until Gen Groves calls releasing him.

(Source: U.S. National Archives. Record Group 200, National Archives Gift Collection, "Diary of Lt. Gen. Leslie R. Groves", Microfilm roll #2.)

This incident was mentioned briefly in a government-sponsored historical study published in 1987. Se Bart Hacker, The Dragons Tail: Radiation Safety in the Manhattan Project 1942-1946 (Berkeley: University of California Press, 1987 pp 104-105). According to this source, the family was named Raitliff; another house was discovered nearby with a couple named Wilson.

The Release

According to http:Hanford-downwinders.tribe.net/thread/eddcoce5-44d6-4c51-, the Hanford Nuclear Site in southeast Washington State released many radioactive substances to the air for more than 40 years. This fact sheet discusses how weather affected where the radioactive materials went and how "hot spots" might have formed. "Hot spot" is a term used to describe an area where the concentration of contaminants is greater than that in the surrounding area. Those contaminants can be either radioactive materials or toxic chemicals. As used here, the term "hot spot" refers only to radioactive materials from Hanford. This publication is intended to serve as an introduction to the subject of weather and hot spots. The Network also offers a publication on the history of Hanford's releases: The Release of Radioactive Materials from Hanford: 1944-1972.

In 1943, the U.S. government chose a location in southeastern Washington State for the Hanford Nuclear Reservation, now known as the Hanford site. The federal government condemned privately held property and moved the residents so it could build plants to make plutonium. Because of the wartime secrecy, few people knew why Hanford was built. It was not until August 6, 1945 when the United States dropped an atomic bomb on Hiroshima Japan, that area residents and most Hanford workers learned that Hanford made plutonium. Hanford had begun making plutonium in September 1944, when the B reactor began operating. Hanford plutonium was used in the first atomic explosion in July 1945 at Alamogordo, New Mexico, and in the bomb that was dropped on Nagasaki, Japan (August 9, 1945.

For more than 40 years, Hanford released radioactive materials into the environment. Although Hanford's mission of making plutonium became public knowledge in 1945, most of the public and some Hanford workers did not know about these releases until 1986. In February 1986, the U.S. Department of Energy, in response to public pressure and a request under the Freedom of Information Act, released 19,000 pages of documents, some dating back to

World War II. The documents revealed that releases of radioactive materials from Hanford had contaminated the air, the Columbia River, and the soil and groundwater. Citizen activists played an important role in making these and other Hanford documents open to the public.

The first eight nuclear reactors at Hanford used large amounts of Columbia River Water to directly cool the reactor cores. These eight reactors were at their highest power between 1955 and 1965. Contamination of the Columbia River was highest during this time. The last of these eight reactors shut down in January 1971. People received exposure from the river in several ways: eating contaminated fish and shellfish, drinking contaminated water, swimming in or boating on the Columbia River downstream from Hanford, or spending time along the river or shoreline. According to Hanford Environmental Dose Reconstruction (HEDR) Project was established to estimate what radiation dose people living near Hanford some time between 1944 and 1992 might have received from releases of radioactive materials. Project Green Run refers to a secret U.S. Air Force Experiment at Hanford that released somewhere between 7,000 and 12,000 curies of iodine-131 to the air on December 2-3, 1949. The experiment was called the Green Run because it involved a processing "run" of uranium fuel that had been cooled for only a short time (16 days), and was, therefore, "green". The reported purpose of the Green Run was to test monitoring equipment the Air Force was developing for its intelligence activities concerning the Soviet Union's nuclear weapons program. The Green Run remained a top government secret until the 1980's when reports were made public in response to Freedom of Information Act requests.

Down Winder's Perspective

Written by a woman who has lived all of her life in Eastern Washington and remembers consuming local milk and produce. Her husband loved to fish the Columbia River downstream from Hanford. Name withheld by request.

BETRAYAL

It's as safe as mother's milk, they'll say
When wanting to assure you that it's O.K.
If mom, downwinder, eats Columbia River's fish,
Or consumes white snow-garden salads on the spot
Then mother's milk can become a deadly lot.

So I fed poison to my nursing son
With radioactive iodine-131
Just because we lived in the wrong place
I maimed my babe for that nuclear race.

In 1985 my husband died quite suddenly. Early in 1986 word got out that radioactive iodine-131 had been released in large amounts by the government just to see what would happen to us down winders from the nuclear plant at Hanford, Washington.

With the injuries from my thyroid cancers and the worry over my husband's bladder and bone cancers, I was very angry and felt betrayed by my government. They used us as guinea pigs but we weren't even that good because the government never followed up to see what did happen to us down winders.

The San Antonio Crash

Exactly thirty days after the mushroom cloud cast its shadow over New Mexico, a disabled alien space craft, crashed in Walnut Creek. It was in this crucible of suspicion and disinterest bred by familiarity that a small contingent of U.S. Army passed almost unnoticed through San Antonio, in mid-to-late August of 1945 on a secret assignment. Little or nothing has been printed about the San Antonio crash, shrouded in the "hush-hush" atmosphere of National Security of the time. But the military detail apparently came from White Sands Proving Grounds to the east where the bomb was exploded. It was a recovery operation destined for the mesquite and greasewood desert west of old U.S. 85, at what is now milepost 139, the San Antonio exit of present day Interstate 25.

Over the course of several days, soldiers in army fatigues loaded the shattered remains of a flying apparatus onto a huge flatbed truck and hauled it away. "That such an operation took place, there is no doubt," insist the two former San Antonio residents, eyewitnesses to the event. Reme Baca and Jose Padilla, then age 7 and 9, secretly watched much of the soldiers "recovery work from a nearby ridge". Our keen interest stemmed from being the first to reach the crash site. What we saw was a long wide gash in the earth, with a manufactured object lying cockeyed and partially buried at the end of it, surrounded by a large field of debris. We believed then, and believe today, that the object was occupied by distinctly non-human life forms which were alive and moving about upon our arrival minutes after the crash.

We reported our findings to Jose's Father, Faustino Padilla, on whose ranch the craft had crashed. Shortly thereafter, Faustino received a military visitor asking for permission to remove it. During their school years, Jose and I, best friends would sometimes whisper about the events of that August, which occurred before any of the other mysterious incidents in New Mexico. But we didn't talk to others about it on the advice of Jose's dad and a State policeman friend. The significance of what we saw, however, grew in our eyes over time as tales of flying saucers multiplied across the country, especially in a ban across central New Mexico. Among the most prominent was Socorro Police Officer Lonnie Zamorra's April 24, 1964 on duty report of a "manned" UFO just south of Socorro, less than ten miles north of heretofore unnoticed 1945 Padilla ranch crash. We were long gone from the area by the time UFO's and flying saucers became news, and while both of us kept up with Socorro County events, we lost contact and never discussed the emerging phenomenon with each other. I moved to Tacoma, Washington, while still in high school and Jose to Roland Heights, California.

Remembering The Story

Then, after more than four decades apart, we met by chance on the Internet while tracking our ancestry. It was then, that our interest in the most intriguing event of our childhood was rekindled. During one of the conversations, we decided to tell our story to veteran news reporter Ben Moffett, a classmate at San Antonio Grade School whom we knew shared our understanding of the culture and ambience of San Antonio in the 1940s and 50s, and who was familiar with the terrain, place names and people. This is our story as told to Moffett.

Looking for a Cow

The pungent but pleasing aroma of greasewood was in the air as we set out on horseback one August morning in 1945 to find a cow that had wandered off to calf. The scent of greasewood, more often called creosote bush today,

34

caught our attention as we moved away from this tiny settlement of San Antonio, on our horses. The Creosote scent is evident only when it is moist, and it's presence on the wind meant rain somewhere nearby.

Lightning bolts could be seen dropping to the ridges above the high Sierras. The wind was stiff and the ground rumbled, loud and close. "This is dangerous weather," Jose warned me, "The kind when accidents happen, including flash floods and forest fires." So, as we worked the draws on the Padilla ranch, we were mindful of flash flooding which might occur in the Walnut or side arroyos, if there were a major thunderstorm upstream.

Gully-washers are not uncommon in late summer in the northern stretches of the Chihuahuan Desert of Central New Mexico, especially along the foothills of the Magdalena Mountains looming to the west. Despite minor perils associated with being away from adults, it was a routine outing for us. It was not odd to see youngsters roam far a field doing chores during the war years. "Could ride before I could walk," Jose used to boast to his friends. "We were expected to do our share of the work."

Hunting down a cow for my dad wasn't a bad job, even in the August heat. At length we moved into terrain that seemed too rough for the horse's hooves, and we decided to tether them, minus their bridles, allowing them to graze, as we needed them for our return trip home. We had no idea where the cow might be, past experience taught us, that when cattle give birth, they like to hide in the tall brush. The terrain was pretty rugged, lots of rocks, so we had to do some climbing. The horses would not have made it up there. This is mountainous area, splinters, lots of cactus and ridges, no flat desert area at all. This was the land where time stood still. Where the Apaches and the Piro Indians roamed, where they built their hideouts, out of piles of rocks and had their wars many years ago, and the terrain hadn't changed much since.

Not far from here was located an old Indian burial ground, guarded by monuments containing petroglyphs that people left us with mysterious images painted or chiseled on surfaces of stone, leaving us mystified about how they chose their camp sites, or petitioned their gods for intervention. "Very few people are aware of this location's existence." We had spotted a Mesquite thicket, a likely place for a wayward cow to give birth, and so we set off across a field of jagged rocks and Cholla Cactus to take a look. As we moved along, grumbling about the thorns, the building thunderheads decided to let go. We took refuge under a ledge above the floodplain, protected somewhat from the lightning strikes that suddenly peppered the area. "Sudden death by lightning," Jose thought to himself.

Around Walnut Creek, lightning is far more dangerous than a rattlesnake bite, and it kills instantly. Jose recalled at least one person who had been struck by lightning in the past few years. The storm quickly passed and as they began to move out, when another brilliant light, accompanied by a crunching rumbling sound that shook the ground around them. It was not like thunder at all. "Not another experiment at Trinity," Jose thought to himself. No, it seemed too close.

It was still fresh in his mind, how his classmates used to refer to him as the "little atomic man", how close he was to the bomb explosion and he survived. The ground shook like an earthquake, then there was like a streak of light, not a straight line, this was very illuminating. It seemed to come from the next canyon, adjacent to Walnut Creek, and as we moved in that direction, we head a cow "moan" in a clump of brush, mesquite brush. We quickened the pace, it was hard on your legs and tough on your feet. Lots of rocks and uphill all the way. Sure enough, it was the Padilla cow, licking a white face calf, what a sight. A quick check revealed the calf to appear healthy and nursing, so we decided to reward ourselves with a small lunch Jose had sacked, a tortilla each, and an apple, washed down with a few swigs of water from a canteen.

As we munched, Jose noticed what seemed like smoke or dust coming from a draw adjacent to Walnut Creek, a main tributary from the mountains to the Rio Grande. Ignoring our task at hand, we headed toward it, and what we saw as we topped a rise "stopped us dead in our tracks", there was a gouge in the earth as long as a football field, and a circular object at the end of it. It was barely visible. It was the color of the old pot my mother was always trying to shine up, a dull metallic color. We moved closer and found the heat from the wreckage and burning greasewood to be intense. We could feel it through the soles of our shoes. It was still humid from the rain, stifling, and it was hard to get close. We retreated briefly to talk things over, cool off, sip from the canteen and collect our nerve, worried there might be casualties in the wreckage.

Then we headed back towards the site. That's when things really got eerie. Waiting for the heat to diminish, we began examining the remnants at the periphery of a huge litter field. I picked up a piece of thin shiny material that reminded me of "the tin foil in the old olive green Phillip Morris cigarette packs" that my mother smoked. It was folded up and lodged underneath a rock, apparently pinned there during the collision. When I freed it, it unfolded all by itself. I refolded it, and it spread itself out again. I put it in my pocket.

Finally we were able to work our way to within yards of the wreckage, fearing the worst and not quite ready for it. I had my hand over my face, peeking through my fingers. Jose, being older, seemed to be able to handle it better. As we approached we thought we saw, yes, definitely did see movement in the main part of the craft. Strange looking creatures were moving around inside. They looked under stress. They moved fast, as if they were able to will themselves from one position to another in an instant. They were shadowy and expressionless, but definitely living beings. I wanted no part of who ever, whatever was inside. Jose wasn't afraid of much, but I told him we should get out of there.

I remember we felt concern for the creatures. They seemed like us—children, not dangerous. But we were scared and exhausted, besides it was getting late. Jose pulled out his binoculars and looked thru them. He could see that the gouge ran into a ridge and then stopped. It really plowed up the ground. There were hardly any vegetation left on the ground at all, it seemed to have been stripped. Now, this wasn't a straight line gouge, it made a slight turn before it stopped. It made a slight turn about the last hundred feet where it turned to the right, Jose remembers. It dug itself into the sandy soil. We could see that there was something there, but the blinding dust and the heat were unbearable.

We found the heat from the wreckage and burning greasewood to be intense. You could feel the heat through the soles of your shoes. It was still humid from the rain, stifling, and it was hard to get close. Jose was worried there might be casualties in the wreckage and he wanted to get closer and go in. About that time, I'd had it. I had seen more than I wanted to see, this was really scary, I was ready to go home. Nothing at this point made sense to me. Sounds, pictures, calendarios, and sounds going thru my mind. It was really hard for me to understand.

Memories of the trinity test came back to Jose. "That was the brightest sun I have ever seen", Jose said to me. Months later, he would learn that it was about two hundred times brighter. Everything seemed out of place, a big gouge in the earth, a silence, an eerie feeling, no sign of birds flying or chirping, everything appeared very still, I just had this feeling, like I was in the movies, looking at different scenes, thru my mind, feeling concern for the creatures.

Thru my own personal round screen, I saw images of tall buildings burning and falling on people and vehicles being demolished. Buildings like I have never seen before. Hearing the suffering, screaming and crying going on. In between this, I could see numbers going across the round screen in my mind. I felt like when you are in a lightning storm, the electricity in the air, trying to pull and push you. "The creatures seemed like us-children, not dangerous," Jose said. "I could see little creatures walking around, moving around, they either had gray white overalls, or they had overalls on and had a hood on or maybe no hair, it was hard to distinguish. If they were wearing overalls, they were skin-tight. They were like staggering around, couldn't seem to hold their balance, all three of them. They seemed to glide or slide from one end to the other. But we were scared and exhausted and I decided we should be getting back to the ranch," Jose said.

There were sounds that continuously came out of the object. High-pitched sounds, like the cry of a wounded jack rabbit, sounds almost like a newborn baby crying, "very moving". Unless you have ever experienced hearing a jackrabbit in pain, you wouldn't understand. It's like a newborn "baby" crying, an experience that stays with you for the rest of your life. We must have there looking at them for around thirty minutes or more, it was hard to tell, didn't have a watch. "Had it not been so late in the day, getting dark, we would have gone in," said Jose.

The sound of somebody injured and in pain would have been a good reason to try and help. The "crying" a real high pitched sound, came every thirty or forty seconds, it would stop and then continue. That someone was in distress? In despair? The creatures continued to dart around in the object. I was thinking, if someone's hurt, we should go on in and help, and at the same time, I remembered something Reme had mentioned. If we don't know what the object is, how can we tell if the occupants aren't dangerous? I couldn't argue with that. We couldn't totally see the creatures, dust and debris obstructed our view. I could see smooth heads.

In the dusk of day, we could see scattered pieces of metal. There was a large piece missing from the object, and several pieces scattered around several hundred feet. We could see part of the object as it was lying cockeyed. An interesting part of the object was what looked like a fin or small dome located on the top end of the object, something of an aerodynamic nature. A piece of the object was torn open. Some distant shrubs interfered with clear observation. We could see their heads. They were shaped like a bug, grayish white. The closest comparison would be to a "Compamocha" the face of a dragon fly kind of a bug that we were familiar with in San Antonio. Jose says he couldn't see any structures or shapes inside the object. We were too far.

The object was about ten to fourteen feet tall, twenty-five to thirty feet long. There were no lights flashing, no whistles blowing. We were not able to determine what part of the object was the front or rear due to it having partly buried itself into the earth. On the way back to the ranch, I kept repeating, "Oh San Antone" an expression used by kids at the time. We finally left, feeling sad and helpless. After all, we were just two kids, and there wasn't much we could do. "Maybe the adults could help," Jose thought. It was dark, and out here, it's like pitch dark. So we back tracked, ignoring the cow and new born calf.

It was after dusk when we climbed on our horses, and a few tears may have been shed on the way home on behalf of the "Hombrecitos", (Little Guys,) without either one of us acknowledging it. But, we know what we saw, don't claim to know what it was, but one of the most unusual things about those visions, is that I was looking at the buildings from up in the air, and the buildings were to the left of me, the vehicles and people were below me, and it was all in color. Black, red, and green vehicles, being crushed by falling buildings. These were very tall modern buildings, with street lights, trees, sidewalks fire engines, buses, police cars with sirens.

It was pitch dark when we reached the Padilla Home. As expected, Jose's dad, Faustino met us the door, worried that something might have happened. "Little did he know, and how would he handle it," Reme thought to himself. Faustino asked about the cow and got a quick report. His eyes brightened up, as he welcomed the good news. Faustino was trying to start a white face herd, and in addition to the normal hardships of running a ranch, he was trying to keep thieves from stealing his cattle. This was one of the reasons for having Jose with Reme's help take inventory and check fences, to maintain some semblance of range security, trying to keep thieves from branding his cattle with their own brand. At that time in New Mexico, you brand it, you own it.

After the report, Jose explained, "And we found something else," and the story poured out, quickly and almost incoherently. "It's kind of hard to explain, but it was long and round, like an avocado, and there was a big gouge in the dirt and there were these "Hombrecitos" (Little guys). Our tale unfolded as Jose's father listened patiently. "They were running back and forth, looking desperate. They were like children. They didn't have hair," Jose said. "Will check it in a day or so," Faustino said, unalarmed and

41

apparently not worried in the least about survivors or medical emergencies. "It must be something the military lost and we shouldn't disturb it. Leave your horse here, Reme, Jose and I will drive you home, since it's so late, and I will explain to your mother, she must be worried. You can get your horse later, Jose will give you a hand with that. I am just so proud of you guys finding the cow and calf. Since you, with Reme's help started riding fences, we haven't lost any more stock. Keep up the good work."

Later that same week, State Policeman Eddie Apodaca, a family friend who had been summons by Faustino arrived at the Padilla home. We directed Apodaca and Jose's dad towards the crash site in two vehicles, a pick up truck, and a state police car. When we could drive no further, we parked and hiked to the hillside where we had initially spotted the wreckage. As we topped the ridge, we noticed the cow and calf had moved on, probably headed for the home pasture. Then we all walked the short distance to the overlook. For a second time, we are dumbfounded. The wreckage is nowhere to be seen. "What could have happened to it?" I asked. "Somebody must have taken it," Jose responded defensively. Apodaca and Faustino stared intently, but un-accusingly at us for a few minutes, trying to understand, and at the same time, trying to figure out what is going on? It seemed like an eternity to us.

Finally Faustino leads us all down the canyon nonetheless, and suddenly, "As if by magic," in Reme's words, the object reappeared. "From the top of the hill, the object may have blended into the surroundings" Reme explained recently. The sun must have been at a different angle, and considering the object had dirt and debris all over it. Who knows where that came from? Apodaca and Faustino led the way towards the object, then climbed inside while Jose and Reme were ordered to stay a short distance away. "I can't see nor hear the Hombrecitos,"

Reme offered. But look at the marks on the ground, like when you drag a rake over it. The huge field of litter appeared to have somewhat been cleaned up. Who tried to clean it, I have no idea. Was it the military, or the occupants? Who knows?

The main body of the object, however, remained in place with odd pieces dangling everywhere. Pieces that looked like the angel hair that we used to decorate our Christmas tree. Now it was time for the adults to lecture Reme and Jose. "Now listen very carefully, you are under oath. Do you understand?" Reme quoted Eddie and Faustino as saying. Reme, your father is a friend of mine, and he just started working for the government. He doesn't need to know anything about this, it might cause him trouble. I have explained it all to your mother, and she understands. You are not to go out and talk to anyone about this, Understood? Jose and Reme respond, "We understand. If anyone asks you if you know anything about this, send them to see me. I also don't want either one of you to get into any trouble."

Faustino also worked for the government at the Bosque Del Apache National Wildlife Refuge, and part of the ranch was on leased federal land. Faustino was a patriotic man and honest to a fault in his dealings with the Federal Government, according to Jose. "The government calls them experimental weather balloons," the state policeman chipped in. "I'm here to help. I'm here to help Faustino work out the recovery with the government. They will want this thing back." But this isn't like the weather balloons we've seen before. They were little, almost like a box kite. "You're right, Reme." "Este es un monstruso," que no Eddie? Faustino states. Yeah, it's big for sure, the state policeman acknowledged. "And the hombrecitos?" Reme persisted. "Maybe you just thought you say them," Faustino said. "Or maybe somebody took them, or they might just have taken off." "While Apodaca and Faustino were in the object, they were always visible to us," Jose said. When they came out of the object, they seemed to have changed their attitude. Different than when we were on the ridge and it seemed as if they doubted us. "They were more serious now," says Reme. The State Policeman says to Faustino, "I would think that the military will want this thing back, however, it seems to me they need to proceed very carefully and quietly." "Que no piensas?"

Government Take Over

They need to consider the impersonal attitude they demonstrated for the ranchers in the development of the bombing range. The State Policeman was probably referring to when the flat expanse of land between the Rio Grande, the Oscura and San Andres Mountains, first settled in the 1870's had and had become home to many large cattle ranches. Some of the ranching families in this area familiar names, such as Bruton, Fite, Harriet, Story, Foster, Bursum, Baca, Martin, McDonald, and Padilla. Between 15,000 and 20,000 head of cattle, sheep and goats were raised each year on these ranches. After the Japanese attack on Pearl Harbor, the Government war machine quickly geared up to unbelievable levels to fight this war. Many ranchers along the Jornada Del Muerto Basin were given until April 1, 1942 to vacate their land. It was needed by the government to develop a bombing range to train military pilots. The ranchers had to move, rent small houses or rooms at their own expense, as there was no government subsistence for ranchers at that time. These were patriotic Americans who were asked to make huge sacrifices for their country, including serving in the military. They were never paid for the land.

Official Military Visit

We didn't hear any sounds anymore, didn't hear the "Hombrecitos" no more. We felt kind of sad. Then we all headed home. The cow and calf also grazed their way back to the ranch house in a day or so. It appears that the official military involvement began after the visit to the crash site by the State Policeman and the group.

After we were done picking vegetables from our family garden, and we had just arrived at Jose's home and it was then that a Latino Sergeant named Avila arrived at the Padilla home in San Antonito, a tiny southern extension of San Antonio. As we drove into the driveway, got out of the pickup carrying two large paper sacks, went to the back of the house and walked in thru the kitchen door, set the bags on the kitchen table and proceeded to join Mr. Padilla who is in the middle of a conversation with the soldier. The soldier stares down at us accompanied by the old man, all the time keeping his eyes alert for any suspicious circumstances in his surroundings, indicating that this soldier had been properly trained.

Come on in, said the old man, unlocking the screen door. The soldier extends his arm offering both the old man and the two boys a friendly handshake. Mr. Padilla, my name is Sergeant Avila. I have been sent here by the United States Army to get your permission to tear down part of the fence adjoining the cattle guard on your property. We need to install a metal gate in order that we may recover one of our experimental weather balloons that has been reported to have accidentally crashed there. Our eyes lit up, remembering when we had first discovered the mysterious object, and couldn't wait until this guy left. Now that the United States Army is interested in it, "It must be important," Reme whispers to Jose. "What have we gotten ourselves into," he thinks to himself. Mr. Padilla scratches his head, looks out the window, puts his thumbs under his

45

suspenders, then turns toward Sergeant Avila and asks, "Why can't you go in thru the cattle guard like the rest of us?" The Soldier nervously clear his throat and rubs his hands before explaining, "The problem, I'm told, is that your cattle guard is about ten feet wide and our tractor trailer is much wider than that. We will probably have to grade in a road in order to do our work, but I'm sure you could a good road. We will build a good gate for you too.

While we recover the downed weather balloon, we will also need a key to your gate to enable us to get in and out until we can install our own. Would that be agreeable?" The rancher shrugs his shoulders, "Bueno" sure, why not." "We need your full cooperation and assistance in this matter, Mr. Padilla," and the soldiers says, "We would ask that you make sure nobody goes to the site, other than those who are authorized by us. You can be a lot of help to us and to our country by making sure that no one knows about this. I'm sure you understand. Someone has already explained this to you and the boys. It's very important." "How long do you intend to work there?" Padilla asks. "Well, were hoping less than ten days. How do you figure on bringing in road building equipment without arousing suspicion from the towns people? They may have a lot of questions. We could use the intentions of grading some of the roads that have been damaged by flood, and the repairing of some old mine shaft in the area." "Good, we are in agreement then," and the soldier smiles, relieved his mission was made easier by the old man's cooperation. "I'll be on my way," the soldier says. Extending his arm in a farewell handshake, he smiles, "Thanks for everything Sir. We'll be in touch soon, and you guys take care now," he says, pointing his finger at the two boys. Sergeant Avila walks out the door, down the steps and into his vehicle, backs out onto the dirt road and heads for the highway.

It wasn't very long after the Sergeants departure that the Army was on the scene with road building equipment. Long before the road grading began, soldiers were at the site carrying scraps of the mangled airship to smaller vehicles that were able to immediately get close to the scene. These appeared to be Army personal. I visited the crash site continuously, some times with or without Reme. Although we were warned by my dad, to stay away from the area, we felt justified in our visits, since the area was part of the ranch, and was included in our daily work schedule. In addition to that, we had been instructed that it was our responsibility to maintain some semblance of security. We sometimes shared a pair of binoculars, watched from hiding as the military graded a road and soldiers prepared for the flatbed's arrival.

Jose came over to my house and got me on various occasions, when we made our inspections and check on the stock. According to Jose, four soldiers were stationed at the wreckage at all times, with shift changes every twelve hours. One soldier stayed at a tent as a guard and listened to the radio. I could sometimes hear western music. The soldiers would throw some of the pieces of material down a crevice. I suppose they didn't want to carry them up the hill. They would kick dirt, throw rocks and brush over them, in order to cover the pieces up, I suppose. The soldiers would work for a while, and then lock the gate, climb in their pickups and go to the Owl Café, where they'd meet up with girls. I was aware of that because one of my (female) cousins who were there told me so.

There was a time, soon after the crash, that we felt we might be discovered. After getting our work done, we visited the site, with the intentions of going in the wreckage. We felt that we might be discovered soon after the crash when we visited the site after getting our work done with the intentions of going in the wreckage. Our plans were interrupted as we arrived. We sat on our horses on the ridge overlooking the crash site, and we discovered some military movement taking place. Some were standing around the wreckage, while others were picking up debris. They didn't seem to stay very long, and they soon drove off. Apparently they didn't see us, or they didn't care, or maybe we did a good job of blending into the tall brush. There was one person that stayed longer than the rest, and by the time he left, it was too dark for us to go in.

On the following days, once the flatbed truck was in place, the soldiers used wenches and a crane to hoist the intact portion of the wreckage in place. Our eyes opened in bewilderment, to discover that on top of that trailer, we could see the part of the object that was not visible to us before, because it had been embedded in the ground. From behind the side of a hill we observed. The soldiers had built an L-shaped frame. They tilted it to get it to fit the wreckage onto the trailer, because it bulged out over on one side. They placed the wreckage on its side. The ripped portion was to the bottom outside of the frame. The fin was very pronounced and thin. Thru the binoculars, we couldn't see any markings on the damaged side. There were three twelve-inch oval shaped orifices on the bottom of the wreckage. They resembled an oval shaped racetrack. This was a smooth surfaced object, the bottom half was darker than the top.

The soldiers used a single cable on the crane, loaded the avocado and leaned it over about eighteen degrees, and we believe the reason for that was because they had to go under the electric, telephone, or telegraph wires in the town. It was now getting darker and the site was lit up. We could see several people moving around, attempting to cover the large craft with canvas and securing the ends to the trailer with pieces of rope. It appeared that the craft would soon be transported. A thought came to my mind.

Every minute we lay there, we were risking detection. Jose motioned for to me follow him. We returned down the hill where we had come from. Jose whispered, "If I'm right, it shouldn't be too long before they complete the task of covering it up. It looks as if they have leveled the ground somewhat. I would imagine that it shouldn't take too long before they leave to see their girl friends at the Owl Bar and Café, maybe then we can go in and take a closer look." He was right.

A little while later, when we returned to the ridge, the place was empty. Everything was gone. What once was a working place became an empty space. "They are probably on their way to the Owl Café," Jose said in a soft voice. Follow me lets go around this way, to where the truck and trailer is parked. He mentioned how the crevice had been filled with dirt and was no longer visible. We walked up along the edge of where the crevice was before it had been filled in. We came to the road outside the gate, where the truck and trailer with the large object stood. It was covered by an olive drab canvas that looked out of place in this rugged desert valley. We approached the trailer from the rear. "So this is what a low boy looks like up close," said Reme. "That's about it," Jose responded. The soldiers are gone. "How about a Souvenir," Reme said. "Great idea," Jose agreed.

We closely examined this long trailer, stepping it out, trying to get an idea of its length. It must have been almost forty feet long. It stood no more than three feet above the ground, with a tall object resting up against a slanted platform the length of the trailer and extending four or five feet the width of it. The object appeared larger now, then when they first saw it. Jose untied some ropes at the rear of the trailer and lifted the canvas, exposing a portion of the outer of a very large grayish white metal object with a smooth surface to the touch of the hands. Jose asks me to hold the canvas so he could climb in. Reme stretched out to pull the rope, exposing more of the object. Jose climbs in and then he comes back out. He goes towards the cab of the truck and comes back with what we called a "cheeter bar", (a piece of iron used to check the tightness of a chain holding down a load on a trailer.) He climbs back in. In the mean time, I'm trying to make sense of the inside. Shreds of debris that look like angel hair that was used during Christmas Time to decorate our Christmas tree hanging throughout the inside of the object.

That fin is something else. I still cannot understand what I'm looking at. I hear something like a crunch, and about the same time, Jose comes out and hands me an odd looking piece of metal, maybe about a foot long. While I examine it, Jose ties down the canvas. The piece of metal seems to remind me of an ancient sundial, a picture that I had seen in some school book. It's very light and cold to the touch. "Let's call it our "Tesoro"," Reme says as he hand him the piece back and Jose wraps it in his jacket. "Let's get out of here before we get in trouble."

We worked our way down the hill, climb on our horses and head for home. It is believed that off the object went, in the middle of the night through San Antonio and presumably to Stallion Site, on what is today the White Sands Missile Range. Was this clandestine operation undertaken to recover an experimental weather balloon, as we contend, or was it something far more mysterious.

Military Returns

Sometime after the recovery had been completed, some soldiers showed up on the Padilla Ranch, searching the road where the object had been loaded on the lowboy. The soldiers brought the road building equipment back and turned over the dirt on the roads they had built.

This went on for a few hours, it appeared that the soldiers were looking for "something", however they did not make it known to Mr. Padilla what that "something" was. While this was going on, Jose brought the "Tesoro" my house, and I buried it under the floor boards in our storage room across the street in the northeast corner of the room, Reme said.

Apparently the soldiers came to Mr. Padilla's house and asked permission to go thru room by room looking for something. The soldiers asked Mr. Padilla if he had anything that might belong to the military. He directed them to a storage room in the back of the house where he kept weather balloons, and odd pieces of metal he had found. The soldiers went through and took everything, including old voter registration papers.

Chapter Three

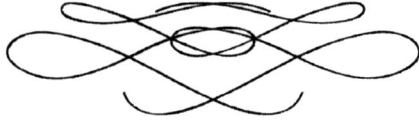

The Sheep Herder

Traditionally, in the fall of the year, sheep would be herded thru San Antonio to the stock yards at the railroad station located just at the edge of town, where they would be kept thru the night and on the following morning they would be loaded on to railroad cars to be transported to the slaughter houses. Mr. Anaya, a lifetime friend of Alejandro, my dad, worked as a sheepherder and while in town, as usual he stopped by and gave my dad a lamb.

My dad was home on vacation and working on different projects around The house. My dad would let Mr. Anaya stay at the vacant house across the street from where we lived, which had a storage and a grain room. Mr. Anaya would live there for a few weeks, usually visiting with friends he hadn't seen for a long time. Some of the kids in the neighborhood and I would congregate on the steps, where he would share some of his experiences working in the wide open spaces of our lands.

One morning in particular, Mr. Anaya came over to the house very early and wanted to talk to my dad. I answered the kitchen door. We were just getting done with breakfast. My dad immediately recognized him and invited him in, offered a cup of coffee, which he accepted. Mr. Anaya was very excited as he proceeded to tell him that last night while he was in bed, he saw a very bright light, and as he looked out the window, it seemed to be by the wood pile behind the house. I kept looking at it, and then right in my bedroom, three little ugly guys, about three feet tall, started putting things in my mind. They were sure ugly looking critters. They were telling me thru my mind, that there was "Tesoro" underneath the floor, and that it belonged to them. "Lo Querren," they want it. I didn't know what they were talking about, so I grabbed my 22 rifle and they disappeared. "Went right thru the adobe wall. I'm going to move out. Yo no se que pensar." "I don't know what to think," Alejandro replies, "Lets go over and have a look."

Alejandro, Reme and one of his brothers joined him and went across the street to the house to take a look. Mr. Anaya, pointed to the area of reference, which was the middle of the room, and Alejandro instructed the boys to pull up some floor boards and take a look. They did and then they also dug underneath the flooring, but came up with nothing but, match books, bubble gum sports cards and trash and debris. After replacing the boards, Alejandro Assured Mr.Anaya that it would probably not happen again.

Later in the week, I met Jose at the Post Office, and told him the whole story, adding how scared I was that they would find our "Tesoro," and how would my mom and I explain the whole incident to my dad. Not good. I had buried it in the North East corner of the room underneath the flooring, where they didn't dig. Jose picked up the "Tesoro," and buried it underneath his own house In 1963, when he had returned to San Antonio to repair his windmill on his ranch, he picked up the "Tesoro" and boxes of odds and ends and took them with him to California, putting them in the attic, where they sat undisturbed for many years. About a week later the old Padilla ranch house in San Antonito burned down.

Applied Alien Technology

Before Jose moved from New Mexico to California, to fulfill his dreams and aspirations, my dad had asked both of us to try and repair his windmill. The pumping cylinder was leaking water, as I remember correctly. The top of the cylinder consists of two parts, the cylinder with outside treads and the cylinder "lid" or top cover with inside treads. The treads were worn out and when you screwed the lid on to the cylinder and tried to tighten it, the treads would slip. If the cylinder was not air tight, it could cause the air to leak in on its upward stroke, and leak out on its downward stroke, preventing a vacuum from forming and that is what would be needed to help pull the water out thru the pipe. We tried everything we could think of, other than replacing the cylinder. "I remember," Jose added. "The cylinder we're referring to was made out of brass.

We took it to Socorro, the County Seat, to try and get it welded. The welder said it couldn't be done because it was made out of brass. I remember climbing down into the well and we had just started putting it back together, when in an act of desperation, Reme asked me to reach back in this wall, behind some salt cedar casing, which kept the dirt walls from caving in. I did and pulled out a red can of Prince Albert pipe tobacco and handed it to him. There were some Indian head coins in that can. Also that is where I kept that piece of Aluminum type foil that I had picked up years before at the crash site. It was a strange piece of material. If you crinkled it up, it would straighten right up."

My dad worked very hard at the Veterans Hospital, nearly a hundred miles away, and he was lucky to afford to come home once a month for a few days. It was important for me to do my best to get that cylinder repaired. Jose and I had no idea that this piece of strange metal had any value at all. We wrapped it on the worn treads and it worked.

The cylinder held vacuum and water came out of the pipe after we primed and turned the windmill on. My dad was very grateful that we had repaired the cylinder. He never forgot. He was always giving me credit for having succeeded where others had failed. Later on that year, Jose and his family moved to California and we communicated by mail for a while. I moved to Tacoma, Washington after having completed my first year of High School in Socorro New Mexico.

Dad' Visit

I remember, that in the mid 1960's, while living in Washington, married and raising a family, my dad came to visit us. My wife, Virginia, was aware that her father-inlaw's favorite dish was lamb chops, so that's what was on the menu. While eating dinner, my dad looked at me and said, "Do you remember that cylinder you and Jose fixed for me on the Windmill?" I replied, "Yes, I did," hoping for the best, yet preparing for the worst. "Well I don't know what you did to it, but I do know it's still working." In a sigh of relief, I thought to myself, if my dad only knew the truth regarding the material we used. Should I tell him? "Alien Technology?" No, he wouldn't understand. We used the only piece of "funny metal" we had recovered from the crash site to repair the windmill cylinder over ten years prior. We were not aware, of how rare that metal might have been. Unknown to anyone, we had saved other material, taken from the crash site.

California Bound

When his family moved to California, Jose assured me that as soon as he got settled down, he would return and take me with, so I could continue his education and work at a part time job. I was looking forward to this, he would be moving to California, same culture, language, and a new set of friends, what an opportunity. It sure sounded exiting.

Unknown to Jose, my brother had separated from the military in Tacoma Washington, went to work there, He asked me if I would like to come and visit for the summer. I accepted the offer and left for Tacoma Washington in the summer of 1955. My visit turned permanent, and I enrolled in High School there. Jose, made a trip to do some repair work on the ranch in the late fall, and stopped at Baca's home to discuss with his parents, regarding my moving to California and living with his family, assuring them that he would continue his education, and he had a part time job already lined up for him. My sister Libby informed Jose that I had moved to Tacoma Washington to live with my brother, and she gave him his address. Jose and I communicated by mail for a while, until we eventually lost track of one another.

New Beginnings

When I left San Antonio, I felt I was well prepared to make my way in the world. I would be attending a school with different kids then I was used to, 99.9% of the students at Stadium High School were white. I had spent all my life, until now, in a world of mixed races, with a majority of them being Latinos. It was natural to feel frightened of a new experience; and quite interesting too.

When I first stepped on the School grounds, I noticed the immensity of the building, when I enter the doors of Stadium High, I immediately see things that I hadn't seen in a school before. Lots of classrooms, offices, football field, Gym, swimming pool, lockers, and a large cafeteria, that served breakfast and lunch, at a very reasonable price. There was also a place set aside where students and teachers could smoke in between class and during lunch breaks. A friendly atmosphere including strict rules and regulations that were understood by students, teachers, and administration. Right away, I realized that stadium high was different. It was located in the north end of Town. I had just never been around a lot of White people before, but I should be able to adjust.

It was 7:00 Am in September, his first day of School, and the streets of Tacoma were still a little dark and foggy. Everyone at his home had left for work. The lights were on at the Baca residence on Monroe Street as I was getting ready for his first day of school. After eating breakfast, I puts on a regular shirt over my T-shirt, then zips my blue windbreaker, hang my book bag over my shoulder where I have placed several "Pee-Chees," (folders) that I had purchased earlier, that were part of the list of supplies that were required by the school. It hadn't been raining thru the night as it did most of the time, but by now the fog had lifted, and it had started to rain again.

Looking like an Artic Explorer, I headed out into the wet streets to begin my long journey to school. I first start off on a Tacoma Transit Bus, which goes to downtown Tacoma where I transfer to another bus that takes me directly to Stadium High School. Once I arrived at Stadium, I find myself in a very different world than what I had been used to. Almost all of the students and teachers and administrators are white, discipline is strict, with harsh punishment for breaking the rules. The first thing that is explained to the students by the Vice-Principal, Mr. Christie, are the rules. The three strikes and you're out rule prevailed. For whatever the reason, you'd be sent to see the principle three times, you would automatically be expelled from school.

A wide variety of Academic classes were available. Under the advice of my counselor, my classes included English Comp, Literature, Civics, History and Typing. All of the classes were tougher than anything I had ever

57

encountered before. Most of my classmates could never understand where New Mexico was and were under the impression that it was a part of the "COUNTRY OF MEXICO." To avoid conflict, I would often refer to my home State as the other "Mexico" and some of his classmates appeared to understand that much better, others thought that perhaps I was an exchange student, and would be returning to my Country of origin soon, or maybe they were just hoping I would.

Working for Change

After graduation, military, marriage, collage and various employment experiences, I became convinced that society had not changed all that much. I had lost my accent, but I couldn't lose my "brown-ness". I became involved with our community, ethnic, veteran and political organizations. I then volunteered my services as an interpreter for the criminal justice system, and hospitals in the City of Tacoma.

In order to become more proficient as an interpreter, I enrolled in college Spanish classes. I became involved with Hispanic, Asian, Black, Gypsy and Native American organizations and served as Vice Chair of both the American G.I. Forum and the Tacoma Urban League.

By invitation, I became involved in the Republican Party, where I participated in various local, state and national political campaigns, taking advantage of the political organizational training the party offered. I was asked by the Office of Governor Evans to chair a welcoming committee for President Ford and First Lady Betty. The entourage included First Lady Betty Ford and her friend Za Za Gabor, Conrad Hilton's former spouse, and many republican officials from throughout the United States. I sat at the table with First Lady Betty Ford and her friend Za Za Gabor at the function, it was an expectation that I did not think I would achieve.

During that time, I was appointed to the Pierce County Advisory Counsel to the Washington State Board Against Discrimination, Port of Tacoma Economic Development Council, and Advisory member to the United States Civil Rights Commission. I was appointed to a Statutory Commission, the Washington State Commission on Mexican American Affairs, where I worked on various projects including Immigration Reform, Migrant Housing and employment issues.

Chapter Four

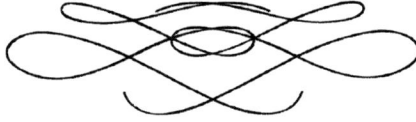

Political Involvement

When Dixy Lee Ray was planning to run in the Primary for Governor, in the Democratic party, after various meetings with her, I joined her campaign, playing a substantial role in the development and coordination of the Hispanic vote in the state wide primary and general elections. Dr. Dixy Lee Ray was a scientist who had achieved high positions in the federal government. She was internationally acclaimed as a Marine Biologist, served as Chairperson of the Atomic Energy Commission under President Richard M. Nixon, and was appointed by President Gerald Ford as Assistant Secretary, Department of State, under Henry Kissinger, becoming the first Director of the Bureau of Oceans and International Environment and Scientific Affairs.

The Nuclear Adventure Begins

The judge's voice intoned her name and ended her daydreaming . . . "Dr. Ray, will you please place your hand on the Bible and repeat your oath after me . . . ?" Dr. James R. Schlesinger was holding the bible. President Richard M. Nixon looked on much too seriously as the routine ceremony ran its course in the President's Office at the White House. It was at that moment that cameras clicked for the TV clips and news photos that would be sent across the nation. The date was August 8, 1972. It was a pleasant event, with smiles all around and an unlimited supply of niceties.

Watergate was barely a whisper and the polls showed Nixon would win in a landslide against George McGovern in the fall. Dixy, unaware of the implications of the recent break-in at Democratic headquarters, as were most Americans, was deeply impressed by the ceremony and the presence of a President. At that moment, she was grateful to be there, a very small part of history. She would do her thing as best she could, then return to her life's work on Puget Sound and enjoy the memories of a brief though perhaps uneventful sojourn with newsmakers. She had no inkling of the political explosion immediately ahead and the path the Watergate eruption would open for her. Schlesinger asked her to be on duty by September 1, for a week of concentrated briefings, so she had three weeks for another fly-back to Seattle and a return across country in her new motor home to visit the nuclear installations.

Dixy acknowledged she had much to learn, but she didn't realize how valuable that trip would until much later. The specially built Dodge "home on wheels" was ready for her when she returned, and the dealer had done a good job. It had to be good. First, it was to make at least four crossings of the United States in the next three years, and second, it was to be her home throughout her stay in the capital. The motor home suited her perfectly. It had ample work space, cleverly designed storage and dressing-room areas, all the bed she would ever want (and a firm, if expensive, mattress—which was fine, because she believed her capability was closely linked to a good night's sleep), compact and remarkable efficient cooking and shower facilities, and room enough to entertain more than a dozen guests—which she did often in the next three years. Big Ghillie, the massive Scottish Deerhound, and little Jacques, the sassy, proud, four-legged Gaul, would be welcome boarders. Ghillie stood as tall as a small horse, while Jacques was a couple barks above a Chihuahua in stature and a thousand in IQ.

Persons who might have come upon the Dixy Lee Ray's motor home just off the rural road might have been startled to see four-star generals and admirals and some of the most important personages in and out of government at a cocktail party, of all things, at the four-wheeled Dodge motor home out in the country. Dixy frequently had VIP's attend her most unusual parties. They didn't mind; why should she? Of all the expenditures Dixy would make in her life, the $21,000-plus she paid for the motor home would be one of her most satisfying. To help her pilot it across the country, she recruited Jane Orr of Annapolis, who had worked for her at the National Science Foundation and whom she quickly hired as her personal secretary for her years in the capital. Jane would become one of her valued confidantes.

Dixy was to become one of the most popular commissioners the AEC had in its three-decade history— popular with the workers and scientists and engineers in the field, that is. The principal reason was the "shakedown" land cruise she undertook before moving to her desk at the capital and her many visits thereafter. She was the first and only commissioner to do it. David Lilienthal, the first AEC Chairman, had visited many of the installations, but he did it as chairman, he did not call on all those Dixy saw. The net result was a educational and morale-building experience on both sides.

Of particular interest to me, was the Dr. J. Robert Oppenheimer, who had been pointed out to me by Mr. Miera, when I was seven years of age and on some occasions when he was real busy, helped him at the Owl Bar and Café in San Antonio New Mexico, prior to the testing of the Atomic Bomb at Trinity. Estanislado Miera stressed that "that man" is going to be very famous in your lifetime. It was under those circumstances that I had asked Dixy about him, as he she had mentioned her escapades in her book, "Is it true what they say about Dixy". Dixy told me that it was described to her by Admiral Lewis L. Strauss, who had been one of the best known and most controversial of AEC Commissioners and Chairmen, as well as friend and confidant to presidents. He had already sent his congratulations in a letter and indicated he would welcome an opportunity to meet and talk with her.

Strauss believed his failure to obtain a degree in physics and the proper PH.D was the principal reason he was rejected, or more correctly, surmised he was rejected by the scientifics' elite. Elitists like Oppenheimer, to be specific. Then she heard the clincher, an explanation of how the Atomic Energy Commission made its decision to create the hydrogen bomb. Now he spoke excitedly, quickly, almost as if the events had taken place that morning instead of a quarter of a century earlier.

"You remember, Dr. Ray, that the vote to go for the hydrogen bomb was a split decision, 3 to 2. What you may not know or remember is that I was quite alone in the beginning as a supporter of the new weapon. It was a very lonely time for me at first, and then others began to see the light. I have not regretted my role in it. In fact, I'm rather proud on recollection. But I shall not forget the bitterness of the time, that awful business involving Dr. Robert Oppenheimer and other scientists." When he said "scientists", Dixy noted a mental wince and the briefest of pauses before and after the word. "I'm sure you know," Strauss continued, "that President Eisenhower and I were very close. We'd been good friends a very long time, but I can say without qualification that I never took advantage of that closeness in government life. When business was being conducted, our friendship marked time. I know many people will not believe that, but it doesn't matter. When we were in session, private, or in groups, on government business, both of us maintained the strictest of official behavior. I never permitted myself to forget that he was the President and I his public servant. That is the way it should have been, and that is the way it was. The Oppenheimer incident was most unfortunate, most unfortunate…that decision caused many heartaches, many cleavages in the nation. Worst of all, it split the scientific community into two distinct camps. It never should have happened. Never. How I wish it could have been averted." He seemed now to be talking to no one in particular and everyone in general as he stroked his chin and looked out into space over her shoulder.

She remembered Dr. J. Robert Oppenheimer and the hydrogen bomb. He had spoken against it from the beginning and Strauss for it from the beginning. Oppenheimer, the organizational genius who had brought so many scientific talents together to usher the world into the Nuclear Age, also had too many friends whose ideological stripe clashed with Red, White, and Blue. He had performed extraordinary service to America, but he hadn't chosen his friends carefully. And who went to his defense when Chairman Strauss and the Atomic Energy Commission declared Oppenheimer a poor security risk and unworthy of handling many of the secrets he had himself discovered or pioneered? Many other scientists.

Strauss had voted to strip Oppenheimer of his security clearance, and, in effect, of his position and reputation. The vote had been 4 to 1, with Commissioner Harry D. Smyth supporting Oppenheimer. The excitement had gone from Strauss' voice. Now he was deliberate, in mental pain. "It was wrong. I was not free to act in that situation. The President—and you know what a stickler he was for strong security—ordered the blank wall," as he put it, "placed between Oppenheimer and all the secrets he had been working for years to develop. Everybody knows about that. But nobody knows he ordered me to remove Oppenheimer's security clearance forever. Mr. Eisenhower was adamant. He demanded that I take appropriate action. So, you see, I was not free to act.."

Dixy was shocked. So Eisenhower forced Strauss to take an action he didn't support, and he knew Strauss would follow orders because his President had so ordered. Dixy remembered the severe criticism Strauss had to endure, primarily from scientists, because he was pegged as the chief villain by those who supported Oppenheimer's position. Dixy herself had been critical of the physicist's bad judgment in playing footsie with known Communists, but she wasn't thrilled by the hysteria that attended his case. Now she watched as a man who had been stamped as Oppenheimer's "executioner" nearly come to tears because a President had ordered him to act in opposition to his conscience. In his book, Men and Decisions, written 12 years before his conversation with Dixy, Strauss had defended his decision to vote against Oppenheimer. Now, to Dixy, he was refuting his own words. He reminded her it was he who had been most responsible for bringing Oppenheimer into the nuclear program and that it was he who had voted twice before to set aside objections to the physicist because he had once associated with Communists.

What would have happened, Dixy pondered, if Strauss had defied the President and backed Oppenheimer, despite the damning evidence? More importantly, what would she have done in his place? Despite the fact that she was never fond of Oppenheimer and his arrogance, she knew. As a scientist herself, Dixy remembered accounts of the merciless way Oppenheimer ridiculed and belittled Strauss in hearings before congressional committees. And now here was the former chairman, who had more reason than anyone to detest Oppenheimer, virtually in tears as he expressed his sorrow for failing to go to the physicist's defense in a crisis.

It was significant to Dixy that throughout the long luncheon, Strauss never made a single derogatory remark about Oppenheimer's behavior or character. A short time after Strauss died the following year, Dixy paid a call on his widow, and Mrs. Strauss corroborated all he had told her. When Dixy considered how long the Strausses had lived with their secret out of respect for the presidency, she was over whelmed with sympathy.

One of the first things Dixy demanded was a breakup of the nuclear-reactor division and separation of safety engineering from reactor development. Her contention from the beginning was that the people who create and develop reactors should not also decide how safe they were. Safety measures, she said, should be virtually in opposition to the development wing. How else could the nation be confident that safety factors would not be swept under the rug? She had the same philosophy in championing the division of the AEC into a regulatory component and a development department.

When she despaired of Washington D.C., she returned to her home, "the real Washington", to run for governor. Her campaign drew little dollar support, so she used her skills as a public speaker and her homespun honesty to shake hands across the state to beat a well-heeled Democrat in the Primary and her Republican opponent in the General Election. "SHE WON". The first Woman governor in the State of Washington. I described Dixy as "GUTSY". She was tough, highly principled, Courageous and brought a refreshingly candid approach to political office.

I went to work for Governor Ray in Olympia, the State Capitol of Washington, where I traveled throughout the state, visiting each of the thirty-nine Counties, enhancing my knowledge of the demographics of each County, working very closely with local, federal and state officials, advising the Governor on various issues, and serving as liaison to community organizations in the development and coordination of Governor Ray's Town Hall Meetings. I attended various functions representing her office, when Governor wasn't able to attend.

After working for Governor Ray, I co-chaired the successful campaign of Doug Sutherland for Mayor for the City of Tacoma. Doug also served as Pierce County Executive, and Land Commissioner for the State of Washington. I served a four-year term as an appointee of Mayor Sutherland on the City Of Tacoma's Housing Authority, and one term as Chairperson. I developed and helped fund the American G.I. Forum's employment program. Serving on the Board of Directors as Vice Chairman of the Tacoma Urban League was another accomplishment of mine.

While my kids were growing up, they were exposed to the political leadership of our local, state and national government. While working in the political system, I felt it important to work with member of either party to accomplish a common objective. A prime example was when the Perce County Republican party chairman Ben Bettridge called and invited me to attend a political forum, during the presidential campaign. He also mentioned that if I had a representative of the Democratic Party's presidential campaign, he would be welcomed to speak on behalf of his candidate. I took Bettridge up on his offer. I only had about 24 hours to arrange this, but I had paid my dues, and Vice President Mondale accepted the invitation.

On the following day, two or three of us accompanied Vice President Mondale to the forum. He was well received. My friend Ben Bettridge's comment was, "You're good." When I had completed my service on the Tacoma Housing Authority Commission, I recommended Bettridge as my replacement. He accepted.

Private Sector

I invested in the plywood industry and served on the Board of Trustees of North Pacific Plywood. After investing in the Insurance Industry, I operated my own agency and also served under the Washington States Insurance Deborah Sean on her Health Care Reform Committee.

"These were the kinds of things we were preoccupied with during the raising of our family," Virginia says. "If nothing else, our family would grow up very much aware of the problems facing our Country and our Society, and hopefully contribute to it solutions."

The Really Big Event

"My husband Reme mentioned his discovery of an object in the New Mexico Desert as a child," Virginia says. "At that time, there was no place for me to fit it into my life or my mind. I rejected it, didn't even want to discuss it. Our children didn't need to hear about it either. They did not want other kids to make fun of them in school. So there it lay. It didn't matter how many times Reme brought the subject up, I would not acknowledge it. Here was no such thing, if there was, God would have let us know through his teachings, after all, I had graduated from St. Leo's High School, and was well versed in God's teachings. In addition, my brother in law was in the Jesuit Order."

"Earthlings were the center of the universe, and that was that. The word unidentified flying objects had no place in our home. One summer, we included my mother and

68

brother on our vacation when we visited Reme's family in San Antonio New Mexico in the mid-sixties, where I learned more about his family and his prior life in San Antonio," Virginia says. "One of the largest Catholic congregations in the county is located here."

It was surprising of the number of adobe houses one encounters in San Antonio in keeping with the truly deep heritage of New Mexico. San Antonio has its share of Frame, Stucco, Brick and other types of houses. The adobe bricks were made right there in town at very little cost. San Antonio is a very small town in Central New Mexico, steeped in traditional, conservative values, beliefs and attitudes of the community. "It seemed to me, that Jose and Reme's discovery of an alien craft, by its very nature would have challenged those conventional beliefs and attitudes. It may not have been acceptable to them either, for it would demand to be confronted, and once it was, it would be impossible to dismiss. So I felt I was on strong footing; we had a great time with our family, and my mother making the comment that if rocks ever became valuable, San Antonio would become very rich."

"It was now 1994, and close to thirty years had passed since Reme's dad had visited us," says Virginia, "and my thinking regarding the existence of Life on other Planets had not changed, preventing me from accepting Reme's experience as a young boy in San Antonio. By now, Reme had made up his mind that it was not all that important for him to share his experience of having discovered the wreckage of an alien craft with society, it was not longer an issue with him. Reme had not been in the UFO Loop, and it would be a monumental task to change my mind," says Virginia.

"That to my surprise came to pass on the evening of July 1994. I, who had become the world's biggest skeptic, until then, refusing to even consider that Reme's childhood tale might just have some substance, found myself under a non-human manufactured extraterrestrial craft. The craft was about three hundred feet above the ground, and about as large as a football field and slowly moving towards and above us, says Virginia. The event took place on a Sunday evening about 10:45 P.M. We were living on South 58th Street in Tacoma, which was located across the Catholic Church, where we attended Services regularly."

"My husband was a Insurance Agent and had an office about half a mile west of the McChord Air Force Gate. He provided insurance Services and sponsored some of their sports teams." Reme was running around in a T-shirt as he usually did when it was warm, and tonight was one of those nights with clear blue skies. South Tacoma way is predominantly a business district, with automobile dealerships, used car lots, restaurants, body and fender shops, auto repair and fast food outlets. The fire Station was located about three blocks from our house. The business district was very lit up to discourage theft, with the Police patrolling it constantly. This was a safe place for us to live. Continuous traffic on the street at all store, café's times of the night. People walking to and from the grocery and restaurants was a normal expectation.

Virginia had studied both music and art in college. She was an accomplished artist and pianist and had taught music for several years. Virginia remembers coming out of the house into the back yard at 10:45 P.M., that evening after getting things picked up in the kitchen. "What are you looking at?" Virginia asked. "I'm looking at the planes coming into the base," Reme answered. Unusual air traffic tonight, as he pointed toward the westward sky. The sky was clear, not a cloud in sight. "See that little dot coming from the bay, way out there, well that going to turn into a plane when it gets closer, that's what I'm looking at, it will probably be landing at the base," Reme says. She joined him in searching the skies and not more than a minute has gone by when she turned to him and said, "Look up above your head." After doing just that, you could hear Reme say, "MY God. Just look at that. It is a huge craft, as large as a football field, very slowly moving towards and above us. It was awesome." "This was not a falling star, swamp gas or a green fireball," Reme said.

The craft came in from the direction of the Narrows Bridge, the bay and made a deliberate turn towards their house. It was now above them. The sky had been covered by a huge craft, and they were underneath it. "Do you know what that is?" Reme said to Virginia. "I'm afraid to ask," was her response. Curious as she was, Virginia describes what seemed to be a probe, a stationary ball light which the craft appeared to follow after making a deliberate right turn, then suddenly came to an almost dead stop, right above their heads. A long silence followed.

"There was this eerie feeling that brought back memories from when I was a Kid," Reme said. "When Jose and I discovered that crashed object near San Antonio, there was a silence in the air. It appeared that the world stood still. Now it seemed to be happening again." "You couldn't find a person to save your soul on the street that night," Virginia said. Dogs quit barking, things didn't seem right. Virginia and Reme are standing there, in their backyard, just fascinated by the craft. Lack of traffic overhead was an exception tonight.

Using her artistic abilities to help her describe the craft in depth, Virginia attempts to do so. "There was this "deep bright orange lighting" that appeared to be imbued in the underside of this giant craft. The lighting seemed to glow rather than shine," she said. There didn't appear to be any light bulbs, Their lighting system must have been in the metal, I would assume. The colors seemed to be projected from inside the craft or emanated from it, instead of reflecting as from a painted surface or a mirror.

"There was no noise at all, no sound. The craft just seemed to sit there as if it had been painted into the sky. It was totally amazing. It was there for way over fifteen minutes, I know because I checked the time when we went back in the house and it was well after eleven. While observing the craft, a thought came to my mind," Virginia says. "I'll go wake up the neighbors, the neighbor lady worked for the fire department. They had bought the house next door and were in the process of remodeling it, so there was a lot of debris in the yard next to us at the time, preventing me from going thru the yard to get to the firepersons house. The craft would probably be gone by the time I got there anyway. Then another thought came to me. What if it decides to land? What will I do? I know, I'll just quickly run into the house if something like that happens."

"They wouldn't take the house, would they?" Virginia and Reme continue to study the craft, it was just so fascinating to them. Reme was thinking that if the Craft decides to land, it would cover the entire area, church, homes and all. One of his former employers had been the Boeing Company, where he had worked on all types of military and commercial air craft, including Air force One. "I had never in my life, seen in any of the engineering drawings, pictures, projections or anything that resembled this," Reme said. "What an experience. Even while in the house during the evenings, we would often hear people talking and walking thru the alley behind our house and the street in front of us. For some reason, tonight was an exception. Then suddenly, the craft began to move, slowly at first, then increasing in speed as it moved away from us, we could see what appeared to be a dome at the top of the craft. As it appeared to pick up speed, it went further and further away until we could see it no longer, so we both went back n the house," Reme said. "I went directly to the bathroom, while Virginia went to the front room where she was laying on the couch when I returned. She was crying and repeating' "God never told us what other worlds he created."

It was after this experience that Virginia began to support Reme's account of what he and his friend Jose had encountered in the New Mexico desert in 1945, when they were just two little kids. While it might have been a coincidence, the name of the church in front of our house is "Visitation" church.

Reporting a UFO

Reme didn't know where to start, in trying to verify what they had experienced. In conversations with military flight personal at McChord Air force base, he was informed that they no longer took reports on UFO's. We just don't pursue or report "Bogies" any more. Reme was referred to an organization, the J. Allen Hynek center, where he spoke with Hynek's son, who took the information and informed Reme that his dad had passed. Reme would soon learn that the subject of UFO's was not at all that socially acceptable. Considering that this would have been Reme's second encounter, he had experienced like zero support of his first one from Virginia. He felt very proud and on top of the world, that he had gained a total number of supporter, "one." But this was a very special one. He realized that he had a long way to go. Seriously think about it. Should you be interested in finding out how committed your friends and relatives are to you, just mention to them that you have just had a UFO encounter, and sit back and observe their physical reactions. Patiently answer their questions and start looking for a new golfing partner. If your friends have had a bad day at work, just mention that you've two encounters. They won't be around to borrow money from you anymore. A word of caution, don't let your co-workers know about your UFO encounters unless you're ready to make a career change. Rumors will fly, and they will stumble over each other to advise your supervisor that because of your encounter, you are no longer capable of counting paper clips. I have come to this conclusion, knowing someone who has seen a UFO, is equivalent to having a reborn experience, it becomes "very Spiritual," however, there's an upside to this, you'll have a lot of people praying for you, and hey, that can't be all that bad.

Relocation

Exactly one year after our encounter, we moved to California, and I became an Insurance Broker, and started a successful business with Insurance outlets in three different cities. I worked in that capacity for several years until we moved back to Washington State in the year 2001 to retire.

Radiation Secrecy

In addition to the responsibility of maintaining the secrecy of the downed alien craft in San Antonio, we acquired other burdens. After the atomic bomb test at Trinity was made public, various Government representatives would talk to the village residents, but it was more in the line of asking questions then providing answers. The reason for rabbits, cows, horses, sheep, goats and other animals growing sore's on their bodies, and going blind were never answered. We were advised that it was ok, not to worry about it.

People that were born with deformities were informed that it was probably due to being born during some phase of the moon. We were not aware or informed that San Antonio was considered Ground Zero, and that according to a report released over fifty years later, we had been exposed to radiation many times over the norm, from ground and rain water, milk, food and the air we breath, perhaps it was due to the bomb test having been carried out under total military emergency.

It had been dirty, and dangerous, however, the residents were uninformed, and we were convinced that our Government would not "lie" to us. How would we have known, we didn't see a Doctor, we were poor people. If you had a toothache, there was always someone in town with a pair of pliers who would pull your tooth and step on your foot at the same time, to ease the pain. If you had a bad burn, rubbing Axel Grease mixed with Pitch from a Cedar tree would work, maybe. That is how it was.

We were trained not to become a burden on anyone. We were taught not to complain about pain, and so we didn't. However, our friends and family members began to get sick and die. There was my uncle who had a sore throat, came to visit us during winter, he had a cough and decided to eat ice from the cement water tank in our back yard. The tank had been filled during the summer of 1945. He died a few weeks later. Must have been something he ate, or something ate him. Then there were newborns, loss of eyesight, extra toes or fingers, or lack thereof. Lung Cancer, Thyroid, Diabetes and Kidney problems. Young people and old people were dying.

As a rule, people lived for a long time in San Antonio. I carried on a conversation with a person that was at least one hundred years old. It may be that moving away from the area and not having participated 100% in 'THE HUMAN RADIATION EXPERIMENTS' may have bought us some time.

Jose and I have both raised a family, his main health issues have been lung surgery, and he has been a non-smoker all of his life, and served in the California highway patrol. My health issues have been heart surgery and Diabetes. We have both tried to make a contribution to society, and maybe this will serve as an explanation to those critics who ask, "Why did you wait so long to tell your story"? As they used to say in the old days. "Ande una milia en mis zappatos" (Walk a mile in my shoes).

Telling my story didn't seem to be a priority, at least not in God's eyes. Helping people was. And I had a choice. I was now married and raising a family. Make it a big issue with my wife. She been brought up a strict Catholic, attended and graduated from Catholic School. Her brother was in the Order.

I started getting involved in community affairs as a result of reading a newspaper article in the Tacoma News Tribune that caught my attention. Apparently a Hispanic lady had been taken to the emergency room at St. Josephs Hospital, but people couldn't communicate with her, so she went untreated and died. The hospital didn't have anyone, or didn't know anyone that could speak Spanish.

I contacted people like David Naranjo, who was working for a multi-service center for the disadvantaged, joined him in his efforts. The lady was survived by two children, if I remember correctly, a boy and girl. We worked with others to get the children adopted. Later on when they were in their teens, they contacted and thanked me for my help. I made myself available to the hospital, informed them that they could call me and I would translate for them, free of charge, and maybe we could save a life or two. They used my services for a long time, I lost a lot of sleep, and missed a few meals in the process, but it was all worth it.

Another case I worked on was the Lantos case. I had gotten a call from a lady, Mr. Lanto's Grandma. She was old and not in the best of health. She stated that her Grandson was on parole, and had just been arrested in Idaho, breaking and entering, and in possession of a weapon. She was aware of my background in working with the disadvantaged. I met with her and she stated that she didn't have much, however she promised she would pray for me for the rest of her life. I couldn't say no. That was her only grandson.

I wanted her to live to see her grandson become a free man, become successful. I wanted her to have great grandchildren. I asked her to start praying as soon as she could, because I was going to need all the help I could get. David worked with me on this case too. Her family paid my airfare, and off to Idaho I went.

I landed in Boise, tried to meet with the prosecuting attorney, but he wouldn't see me. I went to the trial on the following day and wrote a note to the Judge, handing it to the court clerk. The prosecuting attorney was asking for a life sentence, since he believed that Mr. Lantos was a habitual criminal. They were ready to pass sentence asked, if there was a Ringo, Reminko in the court room. I stood up and raised my hand and said, "That's me." He asked me to step up and raise my right hand and be sworn in. He asked me to sit down, identify my self and explain what my interest was in Mr. Lantos. I informed him that I was nobody special, just a member of the community in Tacoma, Washington, and had been asked by Mr. Lantos grandmother to try and save her grandson, so that she might get to have some great grandchildren. This was her only grandson. I had informed her that this was a tall order, but in any event, most of the mountains I had climbed had been fairly steep ones anyway.

I informed the judge that Mrs. Lantos had promised to pray for me for the rest of her life, if I would just try. The Judge asked me, what can you do for this young man? What I said was, we will enroll him in a two year training program as a sewage plant operator, provide counseling and support so that he may become gainfully employed and contribute to our society. The Judge looked at Mr. Lantos:

Would you be willing to follow this mans instructions? His response was, "Yes, your honor." Mr. Baca, I am placing Mr. Lantos on probation to you. If he gets out of line, give the police department a call, and he will be returned to prison for the rest of his life. He will be released to you in about a week, court adjourned.

Mr. Lantos completed the sewage plant Operators class, the transit system hired him and he retired from his civil service position. Mr. Lantos got married, had kids, and went on with life. A few years ago, when I was an insurance agent I was invited to a wedding reception. My wife and I entered, signed in and were directed to a table and sat down. The master of ceremonies called me to the stage, and introduced me to the audience, about two hundred people. This is Ray Baca, a life-time friend of the Lantos family. My grandma told me all about you. You saved my dad's you know what. We owe you big time.

Thank You Reme

By 1945, there was a lot here at trinity to interest any extraterrestrial visitor, including an intriguingly fresh radioactive crater at the site, complete with huge amounts of trinityrite, a product of the freshly melted sand that the bomb test had contributed. Or they might be interested in the contamination of the ground water, and dead animals and vegetation, including the people at ground zero, and those that had lived down wind. Under consideration might have been the long-term health effects of drinking the water, eating the food, breathing the air that had been contaminated by the bomb test. It is reassuring that someone might have been interested, since "OUR GOVERNMENT DID NOT SEEM TO BE, OR WERE THEY? "

Chapter Five

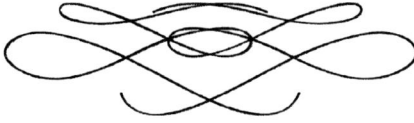

Trinity Test

Yes they were. On the morning of July 16, 1945, the world's first nuclear weapon was tested at the Trinity Site at the northwest corner of the White Sands Missile Range. The atomic blast created a 40,000 foot-high radioactive cloud and a shock wave that shattered windows 120 miles away. Trinity's tremendous light-display was seen as far away as El Paso Texas. Although during the first half-hour the airborne debris from Trinity began drifting with winds blowing towards the northwest, the winds gradually shifted from the southwest (blowing to the northeast). This wind direction would seal the fate of the part of New Mexico extending along Chupadera Mesa and up to 160 miles northeast outwards from the ground -zero- this part of New Mexico extending along Chupadera Mesa and up to 160 miles northeast outwards from the ground zero-this part of New Mexico received one-quarter of Trinity's fallout (with the highest levels extending 100 miles in length and 30 miles in width outwards from the ground-zero).

According to experts who studied Trinity's fallout, 25% of the total radioactive debris landed in this 160 mile elliptical " zone" in northeast New Mexico, and another 25% of the radioactive debris fell at 'larger distances,' which included parts of the heartland (the Plains States) and northeast United States. The last 50% "remained in the cloud in thee form of fine smoke to settle out over a period of months."

Even one full month later a remnant of the radioactive clouds from Trinity-after traveling across the globe-was detected on the second or third day after the Nagasaki A-bomb attack by a U.S. airplane crew at an elevation of 39,000 feet, directly above much of the landmass of the U.S. western coast. The crew described the Trinity remnants as a 'smoke-like layer' that measured 4 to 11 times background radiation levels.

The Fallout of the Fallout

In New Mexico in the days following the Trinity blast, descending fallout debris-described as 'sand-like dust, 'light snow' or 'like…flour' covered the desert landscape. It coated fences and posts, buildings, roofs and cloth lines. It also rained the night after the Trinity blast. At a time in New Mexico history when many ranchers' and farmers roofs diverted rainwater into cisterns and barrels (because a high alkali-mineral content made local ground water undrinkable, or too hard for cooking, or washing ones hair, their water became contaminated with radioactive debris, including plutonium dust.

The rain washed contaminants in great quantities into Tularosa Basin, Pecos Valley and the Rio Grande Valley. The radioactive rain- and the dry, white fallout dust that floated for hours and penetrated into garden-grown foods, cow and goat's milk, wild game and backyard chickens and their eggs. The fallout included Iodine-131, which affects the radiation-sensitive thyroid gland, and about two hundred other short-,medium-and long-lived radioactive isotopes that are nearly all carcinogenic and many considered as immune system-destroyers.

Despite the fact that Manhattan Project and Army staff knew enough, as instructed by their superiors, to evacuate from areas endangered by Trinity's radioactive plumes and follow 'protective measures'-such as closing windows, staying indoors, and even breathing through a slice of bread, the public wasn't given any warning. The public was never given any advice on how to mitigate exposures nor were they monitored for internal doses from radioactive poisons. It was those internal doses to Trinity victims, concluded the authors of a Center for Disease Control (CDC) dose reconstruction report study initiated in 1994, that were never part of scientific and medical evaluations in the years after the Trinity test.

The CDC's 'LAHDRA' (Los Alamos Historical Document Retrieval and Assessment) study, which was completed in 2008 and documented historical releases from nuclear weapons activities in New Mexico with the aid of 'millions of historical documents,' concluded that internal radiation doses via "intake of radioactivity via consumption of water, milk, and homegrown vegetables could have posed significant health risks for individuals exposed after the blast."

According to the 1940 census, 38,000 people lived within sixty miles of Trinity. Most worrisome for the internal exposures of New Mexicans (and others who lived along the path of Trinity's fallout) is one long-term radioisotope called plutonium. The first-generation atom bomb used in Trinity, the "Gadget", required 6 kilograms of plutonium to reach a critical state. This is 150% of the mass needed in present-day nuclear bombs to reach criticality.

Authors Tad Bartimum and Scott McCartney note in their book Trinity's Children that Trinity "was not a terrifically efficient explosion-it didn't use up all of the plutonium in the core. So tiny bits of "unexploded"

81

plutonium was spread over hundreds of miles." The authors explain that "a 1978 inquiry noted a lack of specific information on the plutonium fallout, but said the area was "one of the significant plutonium contaminated areas in the United States, both in terms of quantity of plutonium deposited and area extended"...And a 1983 field investigation noted: "Even after 38 years, there are large areas (near ground zero) whose vegetation is not growing."

Many of the ranchers who lived in the area in the time of Trinity have died from cancer, but no scientific studies were initiated. Unfortunately, the CDC research project (which was funded at a mere fraction of what the present-day Las Alamos National Laboratory institution receives by the taxpayers) never bothered to commence an actual health study of the "down winders" of Trinity. No long-term epidemiological or genetic damage study was initiated. Nor were Trinity down winders the subject of attempts to examine lingering radioactivity in their living environment, or their bodies, or teeth, or urine, or blood.

Some "down winders" of Nevada fallout, to this day, claim they noticeably rise Geiger Counter readings to several times 'background levels' from radio chemicals lodged in their bodies'. Trinity "down winders" alive today may also be hot. Their bodies may contain high levels of radioactive poisons. The same may apply to the deceased.

Determining the primary cause of death for thousands of late New Mexicans may require looking no further than examining the bones of the deceased-plutonium lives on, in their bones. Why has out nation's top health agency made the statement that there ay be 'significant health risks for individuals exposed after the blast' but it is not presently conducting any follow up scientific and medical studies?? It was partly because of the magnitude and distribution of plutonium contamination witnessed by U.S. government scientists and technicians across New Mexico from the "Gadget's" fallout that military strategists weeks later argued for detonating "Fat Man" and "Little Boy" at elevations in Japan high enough to avoid significant local fallout.

Michael Swickard, Ph. D. wrote in his 2010 article 'The real contamination of New Mexico" (NMPoliticis.net) that "When the United States dropped the two devices on Japan it used an "air-burst" method at bout 2,000 feet to keep from really contaminating these areas. While that helped Japan, it was not much help to already contaminated New Mexico." Wash-outs would, however, aggravate the fallout situation following each of the first three atomic detonations on Earth as rains brought radioactivity to the ground, permanently contaminating valleys, northeast of Alamogordo, and the suburbs of Koi, Takasu and Nishiyaman, Japan.

The Trinity Site-and significant surface areas of public and private owned lands to the northeast-is still contaminated by radionuclides from the atomic blast. Some of the worst hotspots-still radioactive today-from Trinity include areas 'around Hoot Owl Canyon and …on Chupadera Mesa'.

In mid-1980s, Chupadera Mesa, which geographically spans the northern-most part of White Sands Missile Range and public areas saddling New Mexico's Route 380, was placed on a shortlist of top"100 sites" under 'FUSRAP's which stands for Formerly Utilized Sites Remedial Action Program. FUSRAP's mission, according to its webpage, is "to identify, investigate, and clean up or control, sites where residual radioactivity exceeding current guidelines remains from the early years of the nation's atomic energy program or other sites assigned to the Department of Energy by Congress."

The Department of Energy, in 1985, conducted radiological survey of the area and 'cleared ' it. The clearance-later summed up in a final report published by Los Alamos Nuclear Laboratory-simply meant that the 'site did not require radiological remedial action,' so no cleanup was done. It turns out that not a single off-site, public area littered by the fallout from continental U.S. nuclear testing has ever been cleaned up.

The Nevada Test Site, which scattered radioactive debris from 100 open-air atom bomb tests and several dozen plutonium dispersal experiments, has raised radiation levels in nearby towns significantly, including Queen City Summit, which is ten miles from the northeast border of the NTS. Queen City Summit has Plutonium concentrations in the topsoil of 0.8 to 1.38 picocuries/gram, or up to two times the expected levels of plutonium found in soil globally from weapons testing fallout and Chernobyl. Several areas to the north and east of the NTS, some populated, according to a 1977 DOE map, contain plutonium concentrations over 80 nanoCuries per cubic meter of soil, or at least five times expected 'background'levels(15.93 nano Curies per cubic meter of soil).

Various areas of Chupadera Mesa that are used for raising cattle and growing alfalfa and row crops are inserting unneed long-term radioactive poisons into the American food supply. These ranchers and farmers-and America-deserve a full-fledged cleanup of their land.

100-Ton Test

Although not entailing a "nuclear device," largely forgotten from history is the fact that an experiment called the "100-Ton Test" preceded "Trinity". It involved the deliberate explosive dispersal of radioactive materials for the purposes of calibrating the project scientists' radioactive equipment. Mixed into chemical high explosives were a small volume of radioactive solution that came from an irradiated uranium slug from Hanford Nuclear Reservation in the State of Washington. The radioactive solution, comprised of fission products similar to that found in spent fuel rods, was as radioactive as the equivalent as all of the strontium-90 –if hypothetically distributed evenly amongst all people in the world-would kill all of the people in the world, of the 1950s, in a few years. This 100-Ton Test concoction-the world's first radioactive dirty bomb-was exploded on May 7, 1945, and dispersed radioactive materials across a large region; about 98% of the radioactive material fell outside the 450 foot radius of the ground-zero, which was about 1 mile away from the Trinity Ground Zero.

Initiator Trials

Manhattan Project scientists, who later became know as Los Alamos scientists-affiliated with the Los Lamos Lab that replaced the "Manhattan District" were busy after the war's end in the PACIFIC. They were behind the scenes of "Operation Crossroads" a series of U.S. atom bomb tests conducted on the now-famous Bikini atoll. Los Alamos scientists then quickly returned back to the Trinity site in New Mexico following these Pacific tests for a new set of atomic experiments. This time they conducted experiments on the "initiator" of the atom bomb, which in its early technological version was a sphere comprised of polonium-210 and beryllium that generated a stream of neutrons when subjected to a chemical explosion. The neutron stream helps "jumpstart" the atomic reaction. The first of these initiator experiments-held in an underground bunker on September 8, 1946-was a misfire, however subsequent initiator tests were "successful." No dated were given for subsequent tests.

The misfired "device was later given the name "Sleeping Beauty." Because of the "remote possibility of an unexpected explosion", scientists made the decision to later excavate the site of "Sleeping Beauty" to "terminate" the combustible materials. They decided to leave the material in the underground chamber for six months to a year before attempting an excavation (Frogman. 1946). Why a 6 to 12 month waiting period was required is unclear but twenty full years came and went and no excavation ever happened. Then in 1967, per the LAHDRA report, because "the Trinity was being considered for national historic registry status, and the number of visitors to the area was expected to increase" meaning the public safety threat from an unexpected explosion loomed even larger-excavation began: 40 feet of sand was removed, the bunker was found, the "failed" switch of the bomb was diagnosed, 100 pounds of chemical explosives were used to "safely detonate the materials and all the equipment in the bunker," and the hole was filled in with dirt. The details of the effort were printed in a 1967 article in the Albuquerque Journal News titled "Sleeping Beauty Detonated at Site By Los Alamos." That excavation, in 1967, was part of a project by Los Alamos called "Operation Sleeping Beauty," which comprised a "health physic" survey and a cleanup of various structures in and around the Trinity Site.

In a nutshell, the "official report" claims that Sleeping Beauty was taken from her catatonic state and "Terminated," but the federal government's story raises some questions.

Firstly, when the British government conducted their own initiator tests in the Australian outback-in 1953-they held them outside. If one considers the carefree attitude of Los Alamos scientists at the time, with their various radiological dispersal experiments, why did they care about a few hundred "curies" of polonium and other short-lived radioactive materials? Why did they conduct this particular set of nuclear material tests underground?

The second question raised by the "official" story is the delay in excavation. We know that tourists converged on the Trinity Site as early as the early 1950s. We also know that the "remote possibility of an unexpected explosion that could cause injury to visitors at the Trinity Site from "Sleeping Beauty" posed a danger through 1967. So, why was the responsible decision made in late 1946 to attempt excavation in six months to a year delayed until 1967. Why were tourists not warned that they could be harmed from a UXO-a land mine at the Trinity Site? The rationale that a new historic status conferred on the "Trinity Site"-bringing more tourists-made the excavation a higher priority implies that Los Alamos officials didn't care about the tourists pre'1967here is no reason given in the official record- the LAHDRA report-for the delay. There is no reason given for why the proposed 1947 excavation was put off for twenty full years,

One can answer these two questions simply-by claiming that safety and / or secrecy required conducting the initiator tests in underground bunkers and that both cleanup and safety even when lack thereof endangered the public-in the nuclear-weapons complex was a low priority during the 1940s, 1950s and 1960s. Public safety, a low priority?

Consider that from the mid-1940s through 1962 Los Alamos scientists conducted secret dispersal experiments of radioactive lanthanum and other materials across the state of New Mexico with zero regard to public health. These radioactive plumes were dropped from airplanes and blown far and wide from explosive open-air tests in northern New Mexico. No warnings were given. In the 1950s and 1960s the radioactive content of foods globally skyrocketed from iodine-131 and strontium-90 fallout from atmospheric nuclear bomb tests (explosions) in the Pacific and Nevada that were planned and frequently executed under the supervision of Los Alamos scientists.

Nuclear hawks would say now that there was only so much they knew back then about radiation, that the health effects are negligible and that sacrifices were need to win the cold war. But the plain truth is they didn't give a shit about people's health. They got away with what they did because they could. So why should we be surprised that they let tourists walk atop a land mine? They got away with it, didn't they? Was "Got away with it" really on the minds of these Los Alamos scientists? Yet, a 2010 article published in the journal Health physics provides a glimpse into the past. It started, in the aftermath of the Trinity bomb test, when a "committee formed in 1941 had voiced concerns regarding long term risks from cumulative radiation exposures of ranchers...although concern was voiced for the health status of at least one family, no evidence was found of steps being taken to reduce exposures to ranchers who continued to live in the fallout zone after July 1945. This was in spite of the fact that soil and the grasses eaten by grazing livestock were particularly radioactive in the area of Hot Canyon. In retrospect, (a group health leader at LANL, Dr. Louis) Hempelmann acknowledged that "a few people were probably overexposed, but they couldn't prove it and we couldn't prove it. So we just assumed we got away with it."

The final quote came from an interview with Dr. Hmplemann in 1986-back in late 1945, shortly after the Trinity test, Hempelmann asked the War Department that the health of persons in certain house near Bingham, N.M. be "discretely investigated." Bingham was the closest town to the north to the Trinity Site, isolated on a stretch of highway between San Antonio to the west and Roswell to the east.

Over the next two years, the residents at "Ratliff ranch" were visited seven times by a team comprised of medical staff, health physicists, LANL scientists and "Army Intelligence Agents" – the residents were given some excuse, a well-thought out pretest, from the visit that would conceal their intention for "casual observation" of "external appearances" and "nonspecific questioning regarding any health complaints". It was a casual guinea pig check-up. The ranchers were never given the real reason for the visits-until sometime in the 1980s. They were never told, "You might want to leave. You might not want to sell that cattle." This was another example of Cold War-era "medical captivity" and negligence.

Who else was part of the environmental and human abuses of the Cold War and got away with it? A staggering number of former and present government scientists, bureaucrats, spokespersons, and high-ranking officials got away with telling us that that "fallout" and terminated" unexploded bombs posed no danger to us. Top medical experts got away with remaining tight-lipped about what they know is "the root cause of our global epidemic-fallout." The jurists and parliamentarians in the highest courts and legislatures got away with barring civilians from suing the government for radiation injuries, citing "sovereign Immunity." All of these entrusted civil servants caught a glimpse into a future when people would learn that their radiation exposures caused their illnesses and their lawsuits would bankrupt the treasures of companies, organizations and entire republics. The only way to "survive" was criminally.

In the days, months and year after thousands of these so-called esteemed, educated, and expert professionals and leaders failed to prevent the wide-spread-and continuing-environmental contamination, thousands of more persons who found out bits and pieces of the truth allowed the crimes to continue by failing to ask questions, to press for justice, to enact change, and to fight to impeach corrupt and lying leaders.

But t didn't stop there. Millions and millions of persons across the globe who had freedoms and the opportunities to act to better the world thought that they didn't need to act. They came to think that the world is big enough that one extra person not doing the right thing will not matter. They thought that someone else would rise up. They felt that with the time and effort they dedicate to raise good children, follow the law, adhere to the teachings of religion, and help charitable causes, "that was all they can do." They didn't care enough to rise up and vote out their leaders whenever they lied, boot out medical experts whenever they skewed findings, impeach their Supreme Court justices whenever they judged unconstitutionally, censor their parliamentarians whenever they passed unlawful laws, or boycott news agencies that censored the truth.

When ordinary citizens succumb to excuses, loyalties, greed, cowardice or laziness to not hold accountable their countrymen and women and government employees when they commit malfeasance, perjury, sabotage, fraud or murder, then clandestine versions of these things-cover ups-easily slither through the holes and cracks and go unchecked, unstopped and unpunished.

In this Swiss-cheese style of self-governance, a cover up that led to billions of persons across the globe who are suffering or are doomed to suffer from the cancer-causing effects of fallout-originated radiobiological poisons managed to glide right through. And, yet, we regard "democracy" and our "civilized society with pride, honor, and patriotism?

Both we and the atomic scientists knew we could get away with not caring, and got away with it. But we really got away with nothing. The scientists, their families and all of our families were bit and got the poison. In a society where no one cares very deeply about anything, anything can and does happen, even the bad stuff. We all got away with not caring so deeply about the world and the end result was that the world ended up not caring so deeply about us.

We are all guilty of omnicide. We all have blood on our hands. And if that is not enough to make us learn our lesson, to see the error of our ways, to realize that citizenship and stewardship are the inescapable duties of the global man and woman, then we have sealed our fate. The deformations of gene pools across biological organisms will spell the end of life. The longer it takes for us to see this truth, the less sand will remain in the hourglass.

Paola, you are so right, the aliens have a message, but society does not wish to hear it. We are experiencing gene mutations in our families, friends and neighbors today. The incidence of cancer is growing world-wide. This may be how our civilization ends. The alien visitations may represent other civilizations preparing to "occupy planet earth" after we are gone.

The Tower

"This material appears to have been subjected to a tremendous amount of heat," the Boeing Engineer said to Reme. He sought John Lester, a Boeing Engineer who was familiar with different types of metal and their composition. This assembly may have been part of an energy drive mechanism. I find the presence of "Carbon Tubes" in aluminum a rather recent high technological development still in its infancy.

Our analysis tends to demonstrate the ability of this metal to transfer heat from one end to the other, thus preventing or retarding meltdown. Remembering Jose's and Reme's speculation that the craft may not have been grounded as it passed close to the radio tower, lightning struck it, thus accounting for the large hole on its side, and the crafts ability to maintain some semblance of control during the crash. "The lightning strike, yes, somehow, that High-voltage jolt may have been responsible for the damage," the Engineer said. "A transference of heat" seems to take place, eliminating the heat friction problem and the resistance to electrical conductivity. Carbon carries electricity much faster than Aluminum. The juice flows faster. This phenomenon is so advanced, so radical, that it represents a potential expansion of conductivity.

"Incredible, isn't it," the Engineer says. "These pieces appear to have been machined. There could be a coating like an M&M candy. It might be composed of Aluminum on the outside and something else on the inside. We realize that finding a coated alloy does not mean the fragments are from an extraterrestrial craft."

Metal Testing

We also realize that aluminum does not corrode, so using it as a metal coating just happens to be part of our twenty-first century industrial aircraft and space technology, so we tend to conclude that it may have been used to build oblong vehicles that traveled through time and space in 1945.

We began working with Dr. Greenwood at the College Metallurgy Department where we used three different scanning electron microscopes to look at the metals and then analyze their elemental make up. The first thing we found was that the metal appears to be an Aluminum Silicate. However, right from the onset, Dr. Greenwood pointed out the unusually percentage of Carbon and was asking how it was possible to do that. Later in the day we cut some samples and polished others in order to take a closer more detailed look and analysis.

We never did a formalized research paper on this, but we did get some great pictures (see Picture: Circuits, Casting Lines, Chapter Nine) of very weird structures in the metals. This is called a "eutectic" metal which means that it solidified at a specific transition phase. Mixture of two or more components in such proportion that their combined melting point is the lowest attainable in some cases as low as 60 degrees Celsius (140 degrees Fahrenheit). They look like little skeletons of bugs squashed into the metal, said the Scientist. Also there are short strands of some other material that we believe may be Carbon Fibers. This sort of "hair-ness" about the metal provoked questions from the academics and electron microscope operators.

Test Verification

Our next step was to get further analysis from a separate laboratory, different people, and different electron microscope, so we used the facilities of another University. I did not have any metallurgists with me this time, only a geneticist but the operator was also a scientist and knew a bit about metals. We took many more photos and analyses, and the outcome confirmed what we had seen in the first results. There is something unusual about that metal.

"The day after Roswell" by Col. Philip J Corso (Ret.), a former Pentagon Official Reveals the U.S. Government's Shocking UFO Cover-up. He tells us how he spearheaded the Army's "Reverse-Engineering Project" that led to today's Fiber Optics, Lasers, Super-Tenacity Fibers and "Seeded" the Alien technology to giants of American Industry.

Chapter Six

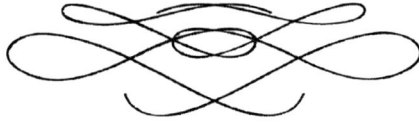

Colonel Corso – They Looked Like Squashed Insect

"I wasn't in Roswell in 1947, nor had I heard any details about the crash at the time because it was kept so rightly under wraps, even within the military. You can easily understand why, though, if you remember, as I do, the Mercury Theater "Was of the Worlds" radio broadcast in 1938 when the entire country panicked at the story of how invaders from Mars landed in Grovers Mill, New Jersey, and began attacking the local populace. The fictionalized eyewitness reports of violence and the inability of our military forces to stop the creatures were graphic. They killed everyone who crossed their path, narrator Orson Wells said into his microphone, as these creatures in their war machines started their march toward New York. The level of terror that Halloween night of the broadcast was so intense and the military so incapable of protecting the local residents that the police were overwhelmed by the phone calls."

"It was as if the whole country had gone crazy and authority itself had started to unravel. Now, in Roswell in 1947, the landing of a flying saucer was no fantasy. It was real, the military wasn't able to prevent it, and this time the authorities didn't want a repeat of "War of the Worlds." So you can see the mentality at work behind the desperate need to keep the story quiet. And this is not to mention military fears at first that the craft might have been an experimental Soviet weapon because it bore a resemblance to some of the German-designed aircraft that had made their appearances near the end of the war, especially the crescent-shaped Horton flying wing. What if the Soviets had developed their own version of this craft?"

"The stories about the Roswell crash vary from one another in the details. Because I wasn't there, I've had to rely on reports of others, even within the military itself. Through the years, I've heard versions of the Roswell story in which campers, an archeological team, or rancher Mac Brazell found the wreckage. I've read military reports about different crashed in different locations in some proximity to the army air field at Roswell like San Agustin and Corona and even different sites close to the town itself. All of the reports were classified, and I did not copy them or retain them for my own records after I left the Army."

"Career after career of anyone in government who even hinted at the big dark secret of Roswell was pulverized by whom ever was behind this operation. And, although I knew far more than I had even admitted to myself, I would never be the one to shoot off my mouth. By now this file, what I would eventually call the "nut File" had come into my possession when I took over the Foreign Technology desk at R&D. My boss, General Trudeau, asked me to use the army's ongoing weapons development and research stream of industrial development through the military defense contracting program."

"Today, items such as lasers, integrated circuitry, fiber-optics networks, accelerated particle-beam devices, and even the Kevlar material in bulletproof vests are all commonplace. Yet the seeds for the development of all of them were found in the crash of the alien craft and turned up in my files fourteen years later. But that's not even the whole story. In those confusing hours after the discovery of the crashed Roswell alien craft, the army determined that in the absence of any other information it had to be an extraterrestrial. Worse, the fact that this craft and other flying saucers had our defense installations under surveillance, and even seemed to evidence a technology we'd see evidenced by the Nazis caused the military to assume these flying saucers had hostile intentions and might have even interfered in human events during the war. We didn't know what the inhabitants of these crafted wanted, but we had to assume from their behavior, especially their interventions in the lives of human beings and the reported cattle mutilations, that they could be potential enemies. That meant that we were facing a far superior power with weapons capable of obliterating us."

"At the same time we were locked in a Cold War with the Soviets and the main land Chinese and were faced with the penetration of our own intelligence agencies by the KGB. The military found itself fighting a two-front war, a war against the Communists who were seeking to undermine our institutions while threatening our allies and, as unbelievable as it sounds, a war against extraterrestrials, who posed an even greater threat than the Communist forces. So we used the Extraterrestrials' own technology against them, feeding it out to our defense contractors and then adapting it for use in space-related defense systems."

"It took us until the 1980's, feeding it out to our defense contractors and then adapting it for use in space-related defense systems. As the ensuing weeks turned into a month, I gradually figured out where some of the puzzle pieces fit. First there are the tiny, clear, single-filament, flexible glasslike wires twisted together through a kind of gray harness as if they were cables going into a junction. They were narrow filaments, thinner than copper wire. As I held the harness of strands up to the light from my desk, I could see an eerie glow coming through them as if they were conducting the faint light and breaking it up into different colors. When the personnel at the retrieval site in the desert outside of Roswell pulled this piece out of the wreckage of the delta-shaped object, they thought it as some sort of wiring device—a harness is what they said— or maybe some of them thought it was a junction box or electrical relay, But whatever they thought it was, they believed there was nothing like it on this planet. As I turned the object over in my hand, I figured, from the way the individual filaments flexed back and forth but didn't break and the way they were able to conduct a light beam along their length, they were a wire of some sort. But for what purpose I didn't have a clue."

"Then there were the thin two-inch-around matte gray oyster cracker-shaped wafers of a material that looked like plastic but had tiny road maps of wires barely raised/etched along the surface. They were the size of a twenty-five cent piece, but the etchings on the surface reminded me of squashed insects with their hundred legs spread out at right angles from a flat body. Some were more rounded or elliptical. It was a circuit—anyone could figure that out by 1961 Col. Corso says, especially when you put it under a magnifying glass—but from the way these wafers were stacked on each other, this was a circuitry unlike any other I'd ever seen. I couldn't figure out how to plug it in and what kind of current it carried, but it was clearly a wire circuitry of a sort that came from a larger board of wafers on board the flying craft."

"My hand shook ever so slightly as I held these pieces, not because they themselves were scary, but because I was awed, just for a few seconds, about the momentous nature of this find. It was like an architectural treasure trove, the discoveries of some long-departed culture, a Rosetta stone, even though whoever crashed onto the desert floor was still very active roaming around our most secret army and air force bases."

"We do want to carry out further tests on the Artifacts recovered from the San Antonio crash, particularly to verify the unusual property this metal has of being able to dissipate heat very efficiently (a bit like the tiles on the space shuttle. It apparently does not melt when subjected to 2000 degree flame from an oxy-acetylene torch for up to two minutes. This despite the fact that aluminum silicate would be expected to melt at about 700 degrees, and this sample, being eutectic in nature, should melt at closer to 500 degrees within seconds. It would not be complex to do these melt tests but we want do them under laboratory conditions, and analyze the metal next with chemical analysis for greater accuracy."

White Sands

Construction began at White Sands proving grounds near Alamogordo New Mexico in June of 1945. Life Magazine published drawings of a manned Space Station as envisioned by the German Rocket Scientists at Peenemunde on July 1945. On August 14th a team of American scientists were dispatched to Europe to collect information and equipment relating to German rocket progress. In September, the Army WAC Corporal's first development flight was fired at white sands proving grounds. It reached a 4,305 mile height using liquid propellant. Jet Propulsion Laboratories) Secretary of War Patterson approved the plan to bring top German scientists under Project Paper Clip to work on missile development at FT. Bliss and White Sands Proving Grounds on October 1945. In December of 1945, more than 100 German rocket scientists and engineers under Project Paper Clip arrived at Ft. Bliss Texas.

Special Research

On January 2, 1946, a special investigation of high temperature aluminum alloys began by J. C. Mcgee, a Field Engineer at Wright Patterson Air Force Base and by June of 1947 had developed a useful alloy known as "MIL" named after the Materials laboratory, which is what the piece of metal that was analyzed and tested by two laboratories was compared to.

Colonel Corso: The Roswell File Artifacts

"There's some more of it down stairs in the file basement that other intelligence agencies don't know anything about. Came here from New Mexico instead of going to Ohio, Don't ask me why," said General Trudeau. "It's coming up to you right now in boxes. Just put everything together, take some time, and evaluate this for me. Anybody know I have this?" I asked. "Everybody knows that if you're poking around something, it's got to be important, so don't act like the cat that ate the canary." "They're watching you as much as they're watching me," General Trudeau responded. Then he walked to the doorway, looked down both ends of the hall and turned back to me. "But move this thing along, because we could be out of this office in under a year and I don't want to have to worry about running out of time on this. And he was gone in a heartbeat, as if we'd never had the conversation."

"I didn't take the file apart that night, even after another nondescript wooden crate that looked like something you ship vegetables in was crated to my office by an equally nondescript army corporal. I didn't go thru the material the next night either. But over the following week, whenever I could be sure that no one was around who could pop in without warning I moved the material from the box into the file and allowed myself time to look at it. It was like falling through the looking glass into a different world, a puzzle of separate pieces that only vaguely captured what had been in the memos I'd read over at the white house. No wonder no one had really wanted anything to do with this junk, which held out the promise of a whole world we knew nothing about but that as far back as 1947, the government had decided to keep an absolute secret."

"Career after career of anyone in government who even hinted at the big dark secret of Roswell was pulverized by whoever was behind the operation. And, although I knew far more than I was willing to admit to my self, I would never be the one to shoot off my mouth. But now this file, what I would eventually call the "nut file" to General Trudeau, had come into my possession, and as the ensuing weeks turned into a month. He describes the devices found aboard the craft, and how they became precursors for today's integrated circuit chips, fiber optics, lasers, night-vision equipment, super tenacity fibers (such as Kevlar Plastic Armor) , and classified discoveries, such as psychotropic devices that can translate human thoughts into signals that control machinery, Stealth aircraft technology, and Star Wars particle-beam devices. He also discusses the role that extraterrestrial technology played in shaping geopolitical policy and events; how it helped the United States surpass the Russians in space; spurred elaborate Army initiativ4es such as SDI (Star Wars Projects) Project Horizon (to place a military base on the Moon, and HARP; and ultimately brought about the end of the cold war.

The medical report revealed that the creatures were enclosed within a one-piece protective covering like a jumpsuit or outer skin in which the atoms were aligned so as to provide a great tensile strength and flexibility. One examiner wrote that it reminded him of a spider's web, which appears very fragile but is, in fact very strong. The unique qualities of a spider web result from the alignment of fibers that provide great tenacity because they're able to stretch under great pressure, yet display a resiliency that allows them to snap back into shape even after the shock of an impact. Similarly, the creature's spacesuits or outer skin appeared to be stretched around it as if it were literally spun over the creature and seized up around it, providing a perfect skin-tight protective fit.

The doctors had never seen anything like it before. I think I finally understood it years later, after I had left the Pentagon and I was buying a Christmas tree. As I stood there in the frosty air, I watched as the young man who prepared the tree for transport inserted it, top first, into a stubby barrel-like devise that automatically spun a twin mesh covering around the branches to keep them in place for the trip home. After I got home I had to cut through the mesh with a knife to remove and separate the tree. This setup reminded me specifically of the medical report on the creature from the Roswell crash, and I imagined that maybe the spinning process of the creature's outer garment resembled something like this.

Chapter Seven

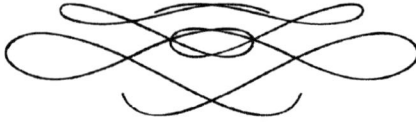

Stallion Site Ranger Center

White Sands Missile Range is so large at 3,200 square miles it takes hours to drive from one end to the other. It is the largest military installation in the United States. It is large enough to contain the States of Delaware and Rhode Island. It can make it difficult for various organizations to support testing missions at the north end of the missile range. Spending three to five hours driving back and forth from a site doesn't leave much time in the work-day to solve this problem. White Sands established Stallion Range Center soon after the missile range opened in 1945. Stallion is located on the northern boundary of White Sands just five miles south of U.S. 380, which connects San Antonio and Carrizozo. It is a 114 mile drive from the main post to Stallion through the missile range. The center consists of a small cluster of buildings that at one time included barracks, a mess hall and a swimming pool and is slowly filling with blowing sand. Security for the north end of the range is headquartered at Stallion. Also, radar and optical instruments used in missile tests are maintained at the range center. The Air force has a presence at Stallion with it ground-based, electro-optical, deep space surveillance system. Known for its acronym GEODSS, IT PRIMARY MISSION IS TO GATHER TRACKING AND IDENTIFICATION DATA ON DEEP SPACE SATELLITES, USUALLY IN ORBITS FROM 3,000 TO

22,000 MILES ABOVE THE EARTH. The system uses powerful telescopes and sensitive cameras to track these satellites. The high desert location at Stallion provides the needed clean air and dark skies required by the system. The White Sands site was the first in chair o these sites now found around the world. The site went into operation in 1982 and is run by contract workers. In addition to the Air Force, another tenant is the Defense Nuclear Agency which recently built the Large Blast Simulator about three miles southwest of Stallion. The simulator uses the explosive release of nitrogen gas to simulate the shock wave of a large explosion. It is the largest shock tube in the world. Just north of Stallion sits a small cluster of hills. One of these hills is known as Compania Hill and is about 20 miles from ground zero at Trinity Site. Many of the VIP visitors who watched the test of the first atomic bomb on July 16th 1945 were stationed on the side of this hill.

New Discovery

Sixty-five years later, Jose's Son was assigned to take some pictures of portions of the Padilla Ranch. As you approach the crash site, it is immediately apparent that something had happened. The area is devoid of Brush. This is as a result of the post impact fire and the radiation. Soil was scorched and, even after nearly 65 years, the soil cannot support significant plant life. A close examination of the site reveals that debris may have been scattered as far as a mile from the impact point.

To our knowledge, there is presently no radioactive contamination in that area of the ranch. Sensitive compounds may have been removed when the gully or Cevice were bulldozed. It is not possible to remove every trace of a crash, no matter how sensitive. Debris can be thrown considerable distances from the impact point. Debris can be buried in soft sand. Debris can be hidden under brush, and searchers have a reluctance to reach in, lest they find a rattlesnake or scorpion has attached themselves to their hand. Debris can be caught in branches while researchers are looking on the ground they may be hanging on brush or cactus in sight. This is exactly how the second piece of metal was discovered, embedded in a dried cactus, in the flight path of the craft, in addition to some other very unusual material that has been placed in a secure vault, until we are able to determine its composition.

The Desert

Hohokam (ha-ho'kaem) is one of the four major prehistoric archaeological Oasisamerica traditions of what is now the American Southwest. Many local residents put the accent on the first syllable (ho'-ho-kahm). Variant spellings in current, official usage include, Hobokam, Huhugam and Huhukam. The culture was differentiated from others in the region in the 1930s by archaeologists Harold S. Galdwin, who applied the existing O'odham term, to classify the remains he was excavating in the Lower Gila Vally. According to the U.S. National Park Service Website, Hohokam is a Pima (O'odham) word used by archaeologists to identify a group of people that lived in the Sonoran Desert of North America.

According to local oral tradition, the Hohokam may be the ancestors of the historic Akimel O'dham and Tohono O'odham peoples in Southern Arizona. Recent work among the Sobaipuri, ancient ancestors of the modern Pima, indicates that Pima groups were present in this region at the end of the Hohokam sequence.

The crevice or Gully that we reference, is where the Natives pitched their tents, watered their horses while on the warpath. There were at one time in that area, burnt hollowed out logs, used as a water troughs, and the arroyos that provided drinking water for their needs. The area where the "Downed Alien craft was discovered in 1945 by the two young boys in the "Chihuahuan Desert." This is a rugged land, and for at least ten thousand years, early hunters and gatherers had crossed these deserts, mountains and high mesas in a constant search for survival. Somewhere back in time, 2000 to three thousand years ago, three distinct Indian societies, emerged from the shadows of the past. The Mogollon who lived in the mountains and high lands of southern Arizona and New Mexico, the Hohokam, Farmer of the southern deserts and the pueblo dwelling Anasazi of the high mesas to the north. These three cultures have had the biggest impact on this part of ancient America. The Mogollon who inhabited this mountainous terrain seemed to have a head start on the road to civilization. Early on there must have been some contact with the advanced cultures to the South in Mexico.

Farmers and Engineers

The Mogollon were the first southwest peoples to use pottery and raise corn. The great achievement of the Hohokam was their engineering. Not only did they build great ball courts and lay out the avenues of their ceremonial cities; they developed systems of canal irrigation, taking the floodwater from the rivers and spreading it out through a series of secondary canals into their fields. Although the Hohokam were still in the stage of horticulture where they worked with mesquite-root digging sticks and seer scapulae hoes, their irrigation systems were far beyond the stage their tools would indicate.

On aerial photographs today the lines of the ditches still show, as straight and true as if surveyed with the most modern engineering equipment. The walls of the ditches were lined with sun-hardened clay to prevent s much seepage as possible. Even before the Spaniards entered the area, Hohokam culture had passed its peak and had

105

descended into a series of scattered horticultural communities, where the people laid great emphasis on funerary rites and observed few other formalized religious practices. It would hardly be an exaggeration to say that the first Spaniards simply walked through these people, hardly noticing them, for detailed written accounts do not begin until the eighteenth century, when missionary activities really begin among the Hohokam descendants. And what was there that the Spaniards would want in their howling desert.

The Mogollon people lefts us with the stunning galleries of mysterious and evocative images painted or chiseled on surfaces of stone. Their clues leave us mystified about how they say their place in the universe, explained the origins of the tribe, chose their village sites in the desert, regarded the epics of their past, commemorated the lives of their ancestors, overcame their anxieties about survival, sought spiritual help for their injured and sick, or petitioned their deities for supernatural intervention or substance. We can imagine the ritual, the ceremony, the celebration, and the dance. Obviously, we have no factual records of the chants of their shamans and medicine men, the stories around their evening campfires, the sounds of their drums. But we do have the stories of their folkway and memories, as passed by elders of the tribes. The stories around their evening campfires, the sounds of their drums.

In the end, like the extraterrestrials, we know very little about the dimensions of their lives. Kay Sutherland, a cultural anthropologist with S.T. Edwards University in Austin Texas, has suggested the Mogollon Shamans may have painted "images" (Archeologists call "painted images") pictographs on stone surfaces and then used the figures as a mystical gateway to the spirit world.

If so the abundance, diversity, distribution and stylistic consistency of much of the Mogollon rock art suggest an intense widespread and pervasive spiritual life, especially in the mountain foothills and desert basins of Southern New Mexico. In any event, if all of the stories and evidence of carved statuary and ruined building is true, America may have the Honor and Prestige of having on its soil, the last survivors in a direct line of the first race of man on North America. Here may have been the beginning of that race, and here may well be the end of that race.

This happy combination, this rare association of the old and the new, this unique blending of the spirit of Ancient Culture with modern progressiveness, may have been the motivation of an Alien Civilizations to explore this land. A Culture that may have at one time been working towards attaining the State of cosmic consciousness, and has now become a warring nation, with the potential power to destroy other planets, including our own.

Chapter Eight

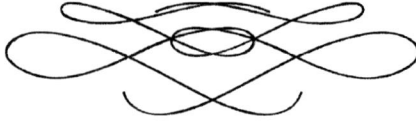

Interview with Reme, Part 1:

SIDE A OF TAPE #1 AND SIDE B
Paola:P
Reme:R

P. After you saw the crash you brought people back there. And who did you bring back? You brought back who?

R: Well not me.

P: Who brought them back?

R: What happened is that after the crash, we went home, back to the ranch.

P: Can you tell me the date of this? The approximate date of this? We know it's 1945.

R: 1945 August.

P: It's in the month of August.

R: And it was like the 15th.

P: Around the 15th. That 15th of August, you is the Feast of the Madonna, it's the greatest feast in Italy. Okay, never mine. It's a major date. Okay, so it's around the 15th so whose dad was it that sent you, and you were how old?

R: I was age 7 and Jose was 9. Jose's dad Faustino had asked us a couple of days earlier, to find a cow that was ready to have a calf.

P: And you were on two separate horses.

R: Yes, we were on two separate horses.

P: So it was during the day you went? Yes. Here's what he told us. You know, Faustino said, when you get a chance, I want you to go out and check that cow because it's getting ready to calve, and we want to make sure that we get it before somebody else gets it, and puts their brand on it. And so you need to, as you get a chance, to go that. What we were doing periodically, is that we'd get on horseback and go up and check all the fences, "riding fence" is what we called it. Check the fences, make sure they weren't broken. If they are, you repair them, you have a small tool kit. If a post is down, you prop it up and later on you come back and replace it. So we'd ride the fences, check the troughs, make sure they aren't leaking, check the windmill, make sure it's pumping water, and its in good repair. So when we were done with that, we'd go up to the top of the hills and take inventory. Jose would look through his binoculars and count the stock I'd write the numbers down. Count the Cows and horses. While twenty-five head of Cattle may not sound like much, Faustino had purchased some white face cattle from Spain, and was in the process of starting a white face herd they seemed to do well in that type of terrain, and brought a good price. They raised two different types of cattle. One was for milk, and the other was for beef. They seemed to well in that type of terrain.

P: Count the cows? Okay. No, believe you.

R: So that's what we'd do. And then of course in the winter when it snowed, we'd sometimes have to break up the ice in the water trough, so that the animals could eat, and transport bales of hay or alfalfa to the windmill area, so the animals could eat.

P: This was during the day that you guys went…

R: Yes., this was during the day. Jose would come over on his horse and we'd saddle mine and we would take off. My mom was aware that we were going to do some work on the Padilla Ranch.

109

P: He was 9 and you were 7.

R: Yes. We went looking for that cow. And so while we were there, it was not abnormal in late summer, to have thunderstorms and lightning and this time was no different so we took refuge under a ledge. Then we continued on. We had dismounted because the terrain was steep and rocky and horses don't do well on rocks, they hurt their hooves. We replaced the bridles with rope and tied them so they could graze and we continued on foot. As we walked towards the clump of mesquite bushes, creosote, or greasewood as they called in the day, pine, sage and cactus. As we walked towards the clump of mesquite we heard a moan and we discovered it was the cow we were looking for, and it had given birth to a calf. This was part of the beginning, of the start of a new herd, a "Whiteface" herd. A red cow with a partially white face and feet. Faustino had purchased a cow and bull, and was breeding them. Whiteface was one of the cattle breeds they used in the United States for meat at that time. So we found it and then we went down into a little area where there was a ledge. Jose had packed a lunch, a couple of tortillas and a couple of apples. We sat down to eat that and the storm and rains came. We got under there so we wouldn't get too wet. Then it just kind of sprinkled a little bit and it was all over. We were getting ready to go up and take another look at the cow and see if it was eating and take a closer look at the calf. While we were doing this we heard this loud bang.

P: You heard the actual crash.

R: We didn't know it was a crash at that time. We heard this sound, like when the bomb went off.

P: The same sound as like when the bomb went off?

R: Similar to same sound as when the bomb went off and it was still fresh in our minds. When the bomb went off, Jose and his mother were up early in the morning. The bomb went off after his dad left for work. Jose's mother looked at the flash thru the crack in the door jab and as a result of the exposure, she lost sight in that eye. According to Jose, they felt the heat wave, and the rumbling of the ground.

P: So the sound was familiar.

R: Very familiar. They were closer to the bomb explosion than I was, my bed crashed against the wall and it bounced me out of it, It was as bright as day, my mother got up and tried to explain hat it was probably that storm that was causing all this.

P: Going back to the actual story, you heard this sound...

R: We heard this sound and the ground shook, and so memories came back of the atomic bomb explosion. Are they testing again or what? So we looked around, saw smoke coming from maybe a couple of canyons down, up that way. So Jose says, let's go over and take a look, see what's going on. We started walking, and we saw a little smoke coming from that direction. As we reached the ridge, the smoke became intense. Then we worked our way down the ridge so we could see what appeared to be a big gouge in the ground. It looked like a road grader had been in there. We were not aware that anyone had a 100-foot wide grader, but it sure looked like a 100-foot wide blade had been here, grading about a foot deep. We started walking up this graded road, it was pretty rough on our feet and it was warm. The bottoms of our feet felt hot.

P: And do you remember around what time it was?

R: I didn't have a Watch.

R: Probably 4 or 5 in the afternoon, maybe later.

P: I'm asking because you can see what you're looking at, it's not dark.

R: No, it's not dark. But as we look up this graded road, there's a lot of smoke. So we retreated, to where we could get some air and take a drink from the canteen and kind of recollect our thoughts and try to understand what this is all about. I asked Jose, is that a plane that went down? I've

111

only seen planes in the air. We live in a small town. Don't see many planes. Jose says, don't know, maybe somebody might be hurt and maybe we need to help them. I said ok, and so we continued trying to get closer. We could see something over on the edge of that graded gash.

P: The path the grader left...

R: The path that the craft left.

R: It doesn't go just straight. It goes and then it makes a right turn, like an "L". We could see something but you know, there's so much dust in the air, and it's humid from the rain and then some of that brush, that oily brush is burning so the smoke's coming into your eyes, it really hard to see and make any sense of it at all. We went back up and rested, returned, and Jose has his binoculars out and starts looking to see what it is. He says, you know there's something over there. Let's see if we can get any closer. Again, we try to get closer and finally it starts clearing up a little. The time seems to be going by very fast. We're looking through the binoculars and I could see the hole on the side of this object. The object is avocado shaped.

P: So it's a round object like an avocado and you could see there's a hole. How far would you say you guys were from the object?

R: I would estimate about a couple of hundred yards.

P: Oh, you did get a couple hundred yards close.

R: About a couple of hundred yards.

P: And then you saw the inside of the hole from the couple of hundred yards?

R: No, not the inside of the hole. Jose says, look at this. So I was looking through the binoculars at these little creatures moving back and forth.

P: Were they moving really fast ?

R: They were "like" sliding.

P: They were sliding...

R: Not sliding, but more like willing themselves from one place to another. That type of sliding.

R: And as I'm looking at that, things began happening to my mind.

P: Oh, really...

R: I'm seeing them and I'm feeling this crazy stuff, like I really feel sorry for them.

P: Um, ,hmm...

R: And I really feel sorry, like they're kids, too.

P0 And you had a concern for them. And you're thinking, did you feel something because of the accident?

R: Yes, I'm hearing this high-pitched sound coming from there. We didn't know what to think. The only high-pitched sounds we were familiar with, were of Jack Rabbits when they in pain, and also the sound that comes out of a new born baby when it cries.

P: I find this interesting. So you heard this same sound...

R: And so that was pretty moving to u. Then I saw these pictures in my head.

P: You did see pictures in your head?

R: Yes, but I didn't know what the heck they were.

P: In other words, you got a telepathic transfer from these beings...you think.

R: Yes, if that's what it was.

P: But you can't remember what they were. But you remember that you got pictures.

R: I can remember what they are, I got pictures, but I didn't know what they meant then, and I still don't know.

P: So they obviously knew you were there.

R: Yes, they must have known we were there.

P: Could they see you if they ever looked out? I mean, no..

R: I don't know..

P: But I mean there was a hole...if hey looked up, could they see these two little boys?

R: Yes, I'm sure they could, if they could see.

P: This was about 200 yards.

R: Yes it was about 200 yards from us. However, there was smoke and dust, so it was not very clear.

P: The beings looked out and they were looking at you, you not only could see them, maybe they were transferring those images to you. So what did you guys do, run away?

R: We looked at them and now it was starting to get dark and we had a long hike to get to the horses and back to the ranch. But Jose wanted to go in and I don't.

P: He wanted to go inside the ship?

R: Jose wanted to go inside the object.

R: And I'm saying, "Jose, what is it?" His response is "I don't know. Okay. If you don't know what the heck this is, I ain't going into it. There's no way. I wanna go home. I don't want to go in. You'll have to go by yourself. I'm going home, I'll meet you at the ranch." And he says, "Well let's watch for a little while. You know, maybe you're right. I don't know what they are. They kind of looked like kids, very strange kids."

P: So you had a whole conversation about this.

R: Oh, yes. And so he says, "Well okay, let's just watch for a little longer and then we need to get back home. Your mom's probably worried, it's getting late, and I'm sure dad's worried..."

P: Can I ask you if you and Jose had a conversation about this, about what you saw, in these years?

R: Jose left San Antonio in 1954 and I left in 1955. During the years we were there, yes we talked about it. From 1955 to 2002 we had no contact. Since 2002, we have compared notes.

P: Have you compared notes? And does he remember things?

R: Better than I do.

P: Better than you do. Oh, good. Okay.

114

R: He has a photographic mind.

P: Okay. All right.

R: He started school at the age of 4.

P: Okay, that's wonderful but we have already got a testimony. So if I was to ask you about the diameter of this thing, how big would you say it was? Did you try to compare to something or...

R: We know. We stepped it off. Remember when we went and pulled that Tesoro off, while the object was loaded on the tractor trailer.

P: Yes.

R: That's when we stepped off.

P: So you stepped it off. What are we looking...

R: About twenty-five or 30 feet long.

P: Thirty feet long. Okay.

R: Fourteen feet high How do I know? Because of the rafters of a house of a house are 14 feet tall.

P: So you got that much information? Okay. So at what point did you guys turn around and walk away? I mean, you were trying to figure out what to do and then what? Did you just turn around?

R: Well, we finally agreed that we ought to go home because it was getting late.

P: Okay.

R: So we started off, went down and got on our horses and started off. It was getting dark then and it was pitch dark by the time we got to the ranch, Jose's dad was waiting for us, he was worried. So we went in and Jose told him the story about the cow and then he started telling him about the crashed object we had discovered.

P: Yes.

R: I told him what I saw and so his dad says, well the first thing we've got to do is we've got to get you home. It seems to me you've had a long hard day. We'll look into this in the next day or so. It probably belongs to the

115

government, and that's probably it. We need to maybe stay away from there, and so they drove me home, I left my horse there. They drove me home and Faustino had a long talk with my mom regarding the object we had discovered on the Padilla ranch. Faustino emphasized it might endanger his job, since my dad worked for the government.

P: Oh, okay.

R: My dad worked for the Veterans Hospital in Albuquerque, and Jose's worked for the Federal Refuge in development, El Bosque Del Apache near San Antonio.

P: Where were they employed before?

R: W.P.A. (Work Projects Administration), C. C. Camps, (Civilian Conservation Corps). They were also employed by Conrad Hilton, who owned several business in San Antonio, and Mr. Alliare, who owned a mercantile business there also.

P: Before that.

R: Yes, before that

P: In their younger days.

Q. And so that was basically what happened that night. On one of the following days, Jose came over to my house ad I went with him to get my horse, and we went to replace a rotted out fence post that we had discovered earlier. That's when we had seen some military picking up stuff, and throwing it into a crevice. Later that week, Jose took me to his house, where we met Eddy Apodaca who was a State Policeman, and a friend of the family. Faustino had asked him to go with us to the crash site, they rode in the state police car, and we rode in the pickup truck. We drove as far as we could get with the vehicles, and we walked the rest of the way to the crash site.

R: When we got close to the crash site, looking down from the hill, we couldn't see the object.

P: What do you mean, did you get very close to the crash site? We are not talking about flat land here. We are

116

talking about hills, canyons, and arroyos. Standing on top of a hill, looking down towards where we had seen the object, it was no longer visible to us, at the time. No explanation why.

R: We simply could not see it. It seemed gone. Jose says well, I don't know what's going on here. Eddy and Faustino said, what did you say you saw? My response was, it's down there, but we can't see it. Faustino said let's walk down there and take a look. We started walking down and then we saw it. The object had a lot of debris over it and so I'm asking Faustino, how come we couldn't se it from up there. His response was that he didn't know.

P: You're saying it was almost it was almost invisible..

R: I almost couldn't see it. Then we got there and they said okay, you guys stay here and we're going to go in.

P: So, Reme, they went in. So what did they find?

R: What ever they found, they did not tell us. What I do know, is they found a complete change of attitude. When we were coming down the hill towards the crashed object, they were doubting us a lot. Almost like as if we had committed a crime.

P: Yes, I know, I know.

R: So they went in and we stood there, sat down on the ground and watched them. And they were in there five or ten minutes and came out. They had a change of attitude, a complete change of attitude. They were almost like different people. They had seen something they'd never seen before. They came out and said, okay. Here the way it is. I want you guys to listen. This is very difficult. "Ustedes Estan Abajo De Juramento" (You're under oath.) You don't tell anybody about this, not your brother, not your cousin, not your mother, not your father, that's our business. We'll take care of that. And the reason for this is that you can get in trouble. We want to keep you out of

117

trouble. So we agreed to that and they gave us a really big lecture, and so we took it very serious.

P: But did they ever tell you what they saw inside?

R: No.

P: They never described it.

R: No. They didn't say what they saw.

P: They didn't. But obviously they didn't see any of the creatures because they weren't there.

R: They didn't seem concerned, because we asked them about the creatures, where are they, because we can't see them through that big hole. There's no creatures there. They said, well, you know, maybe they took off. Maybe somebody took them. Maybe...

P: Was there any evidence the Army had been there, any?

R: Evidence? We saw something like a broom, or rake mark, but then again, it could be some animal, insect or snake that made those marks.

P: Because logically if the military had taken the creatures, they would also have had to show that they had been there in some way. In other words, they waited at least 24 hours before taking the craft.

R: Maybe they did show that they had been there, but we were not aware. Well, before taking the craft?

P: The craft.

R: No, the craft itself took days to get taken out of there.

P: How many days?

R: Oh probably several days. First they would bring in some road building equipment, build a gate, bring in a semi-truck with a low-boy trailer, build a frame on the trailer, bring in a crane and load the craft on to the tractor trailer.

P: Two or three days or more. Did you go back just that once to the area? No, you went back the next day, even though you weren't supposed to.

118

R: We went back several times. Jose went some times with and sometimes without me. You know, we were kids. We worked that area. As kids his dad gives us a little money for doing that work, if we didn't, who would?
P: Had both of you talked about going inside yourselves? Is that why you guys were going back there?
R: Yes And we went there the second day, we were curious.
P: Okay. You were going to go in there.
R: Then we were going to go in there, and we were going to go and see what we could find.
R: We went there on a work day, before Faustino and Apodaca went with us.
R: It was in the afternoon, we were done with our work.
P: Before Apodaca and Faustino went with you.
R: That's right.
P: You went back on your own on the second day. Even...
R: Not on our own, we were working in that area. We had to check that fence too. We had some fences to fix and fence poles to replace. There cattle with calf's around there also.
P: So what happened?
R: Finally, we got there in the late afternoon, we were on horseback and came in from a different direction looking from the opposite side of the ridge, we saw some military people picking up stuff.
P: Okay. Well, that's what I had just asked you before. How did you know the military was there before, you said the creatures weren't there....
R: The military wasn't there all the time.
P: But the creatures were gone and I was wondering, the military must have been there to take them.
R: We did not see the military take them, if they did, it was before we arrived. But we never got to check the object, all we got to do was go down and get some of the debris and

throw it in this crevice and we tried to cover it with dirt and rocks. After the two jeeps left, it was already getting dark and we had to get home.

P: And that's the dig that ultimately someday you want to do. Okay.

R: Yes, that's the one.

P: What did that material feel like, the material that you threw into the trench? Was it like, you know, leaden-like or was it soft or like aluminum, or how was it? Do you have a piece of it? Like stone?

R: Kind of like this piece that I'm holding in my hand.

P: It was like this?

R: It was hard. On the first day, I had gotten a piece of that aluminum foil type, and showed it Jose. It reminded me of the aluminum foil that came in the Philip Morris cigarettes that my mother smoked. I took that and put it in my pocket...

P: Whatever happened to that?

R: I used it to repair the windmill cylinder.

P: So the second day basically you waited until the military went away, and you got more pieces, dragged them into the trench, but you didn't see the beings then.

R: The object was too far from the crevice, almost on the opposite side of the canyon.

R: Too far from the crevice and it was getting dark.

R: The military had been there, we saw them, but I don't think they saw us.

P: The thing was left there and then the next day Jose's father and Apodaca went in.

R: Right.

P: And you took them there, Okay. And then did you see it again? The Object?

R: Yes, it was still there.

P: I mean, you went there a forth day, yourselves?

R: No, no. Then after, probably the third or forth day Jose came over to my house and we picked some Chiles, green peppers, tomatoes, because we had a vegetable garden and they didn't, so filled a couple of bags with vegetables and we took them to his house. We went in the back door, if I remember correctly. And as we pull into the yard, we notice a military vehicle in front and there's a soldier at the screen door talking to his dad, so we go around the back and in through the kitchen. To join them. Faustino says, come on in here boys. So we joined him and he's talking to a Sgt. Avila, and he invites him in. Sgt. Avila says, I'm with the U.S. Army and what I need to do is get permission from you to go in and cut the fence and put in a gate because we have one of our "experimental weather balloons that inadvertently fell on your property.

P: He called it a weather balloon? Those words?

R: An experimental weather balloon, and so we need to recover that, so we need permission to o that. So his dad says, why can't you come in through the existing gate like everybody else does instead of cutting my fence down? Because, he says, the equipment that we're going to bring in is wider than your gate, and it won't fit through there. He says, in the meantime, you have that gate that locks up and we need to have key so we can get in there and cut that fence and put in a gate. He says, we'll put in a good gate for you. And then we need to bring in some road-building equipment, some graders and so forth and see if we can grade a goad o get that truck in and get that weather balloon out of there. So finally Jose's father says, okay, and they both spoke mostly in Spanish. He says, okay, go ahead and do that. Sgt. Avilla says, keep an eye on the place and make sure nobody goes there because you know, this is really important, you know, we don't let any body know about it. We don't want to cause any trouble for anybody, and so try and keep an eye on it, so nobody that hasn't any business going there, doesn't go there. And so, Faustino says, ok, and Sgt. Avilla left and that's when they officially began the process of preparing the area to take the object away. The recovery wasn't anything like what we read about in the UFO books, people in purple uniforms dropping in from helicopters, everything sanitized. It was nothing like that.

P: And they weren't wearing protective clothing____

R: Yes, they were fatigues, they put up a tent, played a radio, western music, and there was always one person left behind in the tent at night when everyone else left.

P: You were watching them, then

R: Yes, we were watching them, as often as we could, sometimes in the morning and evening.

R: If you recall, Sgt. Avilla asked Faustino to keep an eye out, make sure that people who didn't belong there, didn't go there. We were carrying out Sgt. Avilla's wishes, since Faustino was at work, and it was our job to check and maintain the fences, keep track of the herd, including horses. We could hear the music going. There was one guy there at the tent, and two or three sometimes working or picking up debris. They bring in this tractor-trailer, they have a welder, acetylene welder, and they build this rack so they can get the craft on it because it's got to go on sideways. We figured out they were doing that because they had to go under the overpass at about a forty-five degree angle in order to clear it. But then again, these are seven and nine year olds talking.

P: Did they tie it up or put a tarp over it?

R: Yes, put a tarp on it.

P: And tied it up.

R: These soldiers were young men, and they went to the Owl bar and café a lot.

P: Was that the Owl...

R: The Owl Bar and Café, was run by Estanislado Miera, and was located at the crossroads, the junction of 380 East and 85 North and South. In the parking lot, they had a basketball hoop, where we played. They had what they called a fountain where they sold ice cream, shakes and food. They also had a jukebox. So the soldiers socialized between the fountain and the bar. And so we would go there and play hoops, then sometimes Estanislado would come out and ask us to help him. Sometimes we would help grind up meat for hamburgers, wash dishes, clean up the place. This is the famous Owl Bar and Café that is mentioned in the Manhattan Project, referred to as the "watering hole" where physicists like Oppenheimer, Von Newman, and others would stop for lunch or dinner on their way to Trinity, after a long hard drive from Los Alamos. It has been reported that overnight Accommodations were sometimes provided for them during their long journeys. In the days prior to the atomic bomb test, large and small tents were set up in the vacant lots behind the Owl Bar and Café. In the late morning of the bomb test, all the tents disappeared.

P: And so these guys went there.

R: And yes, that's where the soldiers went for lunch and dinner. One of the soldiers stayed back in the tent, when the others were gone. The Object was never left alone.

P: And you saw them pick up debris at the crash site.

R: Yes.

P: They left the ship. And explain to me how you got this metal.

R: On the final day when they brought in a small crane, about I imagine a 15-20 foot crane and they dragged the craft onto the tractor trailer.

P: Did they ever see you?

R: I don't know if they ever did, or cared.

P: In other words, you were like part of the scenery.

R: Well, you know, they weren't looking for us, there was vegetation of the side of the hills, and we weren't very tall, so it was easy for us to hide. But you didn't go and talk to them or anything. What Language would we use?

R: Oh, we would sometimes say hi. but not much, because we didn't have anything in common. The work they were doing didn't seem all that important to them at the time. It didn't seem to be a great deal to them. We don't believe anyone was aware of how important this object might have been, certainly not us..

R: Years later, one of the soldiers married Jose's cousin.

P: You just said one of them married Jose's cousin and the obvious question everybody would be, did this military man who married Jose's cousin every talk about this incident?

R: With Jose?

P: No

R: With Jose's dad?

P: With Jose's dad, this military guy did. Do you remember what was said?

R: I was not there. But Jose would know.

R: It is my understanding that throughout the years, he became more unconvinced that it was a weather balloon.

P: That's what he said, he did not think it was a weather balloon. But he never went one step further and said what was inside.

R: I don't believe he knew. He was just doing his job, picking up the debris, looking forward to completing his assignment and going home. The war had ended, and a lot of the soldiers had been restricted to stallion site for the last 90 days, while the atomic bomb test was carried out.

P: He didn't know. So his job was to do the recovery. But he thought it was not a weather balloon. Okay. So, anyway, you guys were at the fountain and then what all happens with...

R: We'd go there, buy a coke, and listen to the music. It seemed that the guys were not even aware that we existed. They seemed predictable. We pretty well thought we had it all figured out, and on the last day, Jose came and got me and we went to the site, sitting in the brush where they can't see us. We watch them drive the truck outside the gate and they got the tarp tied up nice and neat. Jose said, I think they are going to take it tonight. I said, yea, how about a souvenir?

P: And you said that.

R: Yes, during the war we lost so many relatives that it was not unusual to have something to remember them by when we said our prayers for them. Because they died in the war, they never came back. The military brought a lead coffin and a couple of guards with it and they buried them. Jose say's let's head down and wait a little while until they leave and then we'll go. We waited for a while and then everybody took off. They had these military pickups and they took off. So we know where they're going, they'd be gone for a while. We worked our way up there and where the crevice was, they had run a grader through it, so nobody would even know that a crevice existed. Then we worked our way outside the fence, towards the back of the truck and stepped it off. If you took a big enough step it was three feet. Maybe we were off a few feet, but those are the measurements that we had at that time. 25 to 30 feet long and about 14 feet tall. And then we looked at the underneath part of the craft, because we had not seen this part of it, it was partly under ground. So now we get to see the whole thing. Boy, this thing is a monster, it's big. Now we can see the bottom, and in the bottom, it's got like three little indentations, little grooves under there, on each side.

We at that time had no idea what those grooves represented. We still don't.

P: Well, maybe they were for the landing pad. Maybe some kind of feet came out of it.

R: Could be.

R: And so Jose pulls part of the tarp off, exposing the gash on the side of the craft, while I hold the tarp open. Jose climbs into the Gash.

P: Inside the hole?

R: Inside the hole.

P: He went inside the hole? Yes, and I was partially in, holding the tarp letting the light in.

R: Yes. First, there's nothing hardly in there...

P: But he could see the shape of it? Like if there was any rooms, was it smooth all the way around, if there were panels, if there ...Try to explain it to me.

R: Jose said there were like ridges every so many feet.

P: Did he see any panels, like control panels?

R: No.

R: He didn't see like a panel, like this big of a panel..

R: No.

R: Were talking maybe about two and one half feet long.

P: Was it attached to the wall, this panel?

R: To the bulkhead, the rear wall, maybe?

P: This is on the panel which is inside, which is you said, bulkhead. But it's against the wall, the Panel. NO?

R: Yes, to the bulkhead.

P: Okay.

R: What would be the rear wall to us...

P: So how fast could he pull that thing off? I mean, did he pull it off?

R: He tried to jerk it off and he couldn't so then he went and got a cheater bar from the front of the tractor trailer.

R: Something like a crow bar, its called a cheater bar in the trucking industry, its used for testing the tightness of the chains holding the load down on the tractor trailer.

P: You described the pins and what they were like...

R: Yes, a one-way fastener.

R: They go in one way and they can't be pulled out. The inside of the hole they are in is serrated.

P: They were serrated fasteners that was holding this bracket looking type assembly on the panel that was located on the bulkhead (rear wall.)

R: The pins were like yellow.

P: The pins were like yellow? That was my next question. What kind of color did you have?

What kind of colors are we dealing with?

R: Yellow.

P: The pins were yellow. Did you see any other colored stuff in there? Silvery colored strands of what I would compare to angel hair?

.P: No seats or anything, nothing...it must have been cleaned out, or maybe there weren't any.

Couldn't see any instruments, like gages, clocks, steering wheel, brake pedals, nothing like that.

P: Was it gray inside?

R: Part of that ship was darker on the bottom part than the top. Lighter gray.

P: But did he like race out of there because he thought they'd discover him, or were you guys relaxed because you knew where the military were and they were going to take their time so you just took your time.

R: We tried to hurry, we were running out of light and were afraid of being discovered. Relax? You gotta be kidding. I haven't relaxed since then.

P: This is pretty heavy metal, though...not really, no? Did it feel like an earthly metal? You couldn't tell. But the piece like the Phillip Morris package was different. Where did that come from?

R: When it first crashed and we first went in to the crash site, there were some pieces of material that looked like angel hair. What we called angel hair was used to decorate Christmas trees in that era, since most people didn't have electricity in San Antonio. That material was similar angel hair. I also found a piece of shinny metal.

P: That was moving back and forth...

R: Under a stone, is where I first saw it,. I pulled it out, it had its own mind, I folder it to put it in my pocket, and it unfolded, kind of unusual.

P: How much of that stuff did you guys throw in the trench?

P: That's the stuff that I'm really curious about.

R: That's the stuff everybody wants to get their hands on. We understand. People will do anything to get their hands on that.

P: Sure. Well, that's very interesting. That's what they say the Roswell pieces were like, what Jesse Marcel, Jr., said that his father brought home. He strewed it all over the kitchen floor. There was a lot of it.

P: Were there I-beams?

R: I would not have known what an "eye" beam looked like at the time.

P: You didn't see any structural beams? Not that I remember. So you got to go in it, too. So both of you went in it. So how long did you stay in there?

R: I don't know. Not very long.

R: Here's what was on the craft, when Jose first went in, and after he got the cheater bar, he needed more light, so I climbed partially into the opening., holding the tarp as high as I could.

We didn't have electricity at our house, so when Christmas came around, we decorated the tree with non-electric decorations like pop corn, icicles, foil and angle hair. That year we had a few strands of angel hair that came from a crashed object on the Padilla ranch. You know what that is?

P: Yes, I know exactly what that is.

R: That's what we had...That's what was all over the inside of that craft.

P: Okay. I got another story once and _____told me that that stuff was burned fiber optics. It was fiber optics. It was whitish angel hair. Perfect. Because that's part of the mechanism of the craft. And Charles Hall told me that he also saw a craft with a hole in it that was filled with......

END OF SIDE A OF TAPE #1

BEGINNING OF SIDE B OF TAPE #1

P: So you wanted a piece of the metal.

R: And so we took that.

P: Did you both ever, when you were discussing as kids, did you ever discuss the beings?

R: Yes. Their heads were comparable to a "Compamocha". That's what we saw.

P: What do you mean, you saw them and they looked like a bug?

R: Yes, they were ugly to us at first. Their heads looked like a campamocha.

P: Would you say it in English.

R: The closest translation would be, heck a bug, the praying mantis.

P: Oh, that would have been...

R: Big, bulgy eyes...you know. Everybody calls them grays, I guess, but I havn't seen a gray, so I would not know.

P: But these could have been a total other thing...

129

R: They had big bulgy eyes, we don't know whether they were exactly four feet tall, it's just an estimate.

P: Four foot tall, and they were real thin, need-thin arms.

R: I don't know about "needle thin", and I don't know how many fingers.

P: But I mean they seemed to glide. Were they wearing outfits or...

R: Well, either they were wearing real tight coveralls, or their skin was real tight.

P: What color, still gray, the coveralls?

R: Yes, light gray.

P: And the head was pretty big? I mean proportionally.

R: The head seemed pretty gig, and it was similar to a campamocha.

P: Okay, that's okay.

R: Not like they say...

P: No, no I understand. No, no, no, because I've heard this before it's okay. It protruded, right?

So it was a protruding head and big eyes. And kind of a slit for a nose...You didn't even notice the nose because the eyes were so big. There were thinly, you know, thinly means skinny, it was thinly built.

R: I you see one, I'll have to get a picture of one, but campomocha describes it well.

P: And you said they slid instead of walking or running. They "seemed" to slide.

R: It seemed like they did. Like they willed themselves from one place to another.

P: But you knew something, they must have connected with you at some point. Well you said you had images coming in our head.

R: Yes, I'm sure of that.

P: You know, I wouldn't see eyes unless eyes were looking at me.

R: Jose and I were looking at the craft thru one set of binoculars, we were taking turns.

R: He was looking, but we couldn't look into their eyes, that I can remember, it's pretty far.

R: I know, but what we felt was this pure sorrow, really felt sorry for them because we could feel their pain. They seemed like us, children.

P: Oh, okay. That was certainly interesting. Reme, I have no words for that, to compare something like that.

R: They seemed like they were hurt.

P: They were hurt and they also knew they were looking at your eyes. Anything else that you and Jose, did you talk about them at all?

R: Yes. Did we get together and discuss this when we were little, you mean?

P: Yes:

R: Yes, we talked about them, when we were sure no one else was around. P: Them, the beings.

R: The creatures. Did they talk to us?

P: No, what did Jose say about the beings? I know how you felt about them.

R: the same.

P: Yes, did he feel sorry for them?

R: Not as much as I did, but he did.

P: Were you terrified when you looked at them or did you want to get closer, or did they just disgust you, or you jut felt sorry for them, or...

R: Normally, I would feel sorry for friends, relatives if something happened to them. I didn't know these creatures. We were curious. They were strangers, we didn't know who they were, but we knew they were different.

P: Oh, okay. So you felt their emotion.

R: That's right.

P: Oh, my. You felt their emotion.

R: And so those sounds, we tried to figure out what the sounds were. We attributed them to be coming from the.

P: That's probably where they were coming from. How long you think that experience lasted, when you were doing that?

R: All the time they were there.

P: Which was?

R: Probably a half-hour to 45 minutes.

P: You stayed a half hour to 45 minutes where those beings were? You weren't scared?

R: I was scared, yeah

P: And you still stayed a...

R: We still stayed there.

R: Jose was curious about the creatures too, he wanted to help them.

R: Yeah, Jose tried to talk me into going into the object ot help them, and I'm trying to avoid it, yet I'm concerned too.

P: Jose was going to go and...

R: We don't know what the heck they are, who they are, and what they are doing there. I don't feel good about it.

P: You're not quite sure what that experience produced...Well I can see that you could be confused about... that's a long time. I mean, I think regular people would get scared and run away from them, but you stayed an...

R: Something seemed to keep us there.

P: Something kept you there. Because you were trying to figure it out.

R.: Yes, trying to figure it out. So then eventually we had to leave. We had to go back to the ranch.

P: You didn't notice any writing or hieroglyphics on the outfits. Too far to see. You saw them "sliding" back and forth and how many again, three to four.

R: Yes, three to four.

P: Nobody heard anything, you just heard this high-pitched...well, their craft was crashed.

R: It was still smoky in the area and it was partially buried.

P: So you got the piece and then who kept it?

R: Jose kept it for probably a few days and then after that he brought it to me and I hid it under the floorboards at our storage place across the street. Jose told me that some soldiers had contacted his dad, and wanted permission to look thru his tool shed and his house, and Jose didn't want to get his dad into any trouble.

P: And then what?

R: The military went thru the shed and his house, they took metal, weather balloons, and voter registration material that he had stored there. R: Then the sheepherder, a long time friend of my dad's came into town as part of the crew herding the sheep to the stockyards where they were kept overnight and loaded into railroad cars the nest day and shipped out. As a matter of fact, we went with him over to the stockyards, where they camped out overnight. They used to make such good soup and shepherds bread, and we were invited to stay for dinner, and then we came home. On the following day, the sheepherder moved in the storage house, and gave my dad a young lamb, which he did every year when he came by. When Jose and I had pulled that piece of that craft, our souvenir, we ha named it "Tesoro." We were the only that knew its name. Translated it means "Treasure."

P: Okay. Treasure.

R: So that was our tesoro. The sheepherder comes over to the house one morning while we're just finishing breakfast. My dad's home on vacation, and he is not aware of our tightly held secret. The sheepherder knocks on the door, I answer it and he says, can I talk to your dad?

Sure, come on in, my dad says, come on in Pedro, lets have a cup of coffee, we're just finishing up our breakfast. So,

133

we're sitting, finishing up and so he says, Alejandro, he says, you know, I'M going to have to leave this place. My dad asks, Why? Well, he says, you know, last night I was asleep and I got woke up. I saw this light out there by the well. There was this light out there and I...

P: Who is this man?

R: He's Pedro, the sheepherder.

P: The sheepherder. Okay.

R: A good friend of my dad's.

P: Okay, okay. He saw a light by the well.

R: By the well. And he says, I looked out the window and the next thing ...there's three critters in my room, in my house, and the door's locked. And so he pointed towards the floor
And they're saying "Tesoro".

P: Oh, no. You never told me this part of the story. This is incredible. Oh, my Lord. Okay, and...

R: They're pointing there. And so he says, Tesoro, there's treasure down there.

R: And so these guys are doing that, Pedro says, and I got my rifle and I'm going to shoot them because they have no business in my house. And so I got my rifle and they're gone.

R: But you know what? They went right through the wall. "Can you believe that", Alejandro?

R: My heart is pounding, and I am silently praying, I don't want to get in trouble. And my dad doesn't know anything about the Padilla ranch experience.

P: Okay.

R: So my dad says, alright. He says to my older brother, let's go over and check, bring a shovel and a crowbar..

P: The eldest brother of..

R: Mine.

P: Yours. Okay.

R: And so he gets the crowbar and undoes some floorboards and he steps down and says, where? Pedro pointed right there.. Right in the center of the room. I'M silently praying, oh God, I hope they don't find it. So he digs in the center of the room and he can find anything, and he digs around with the shovel and there's nothing. He says, there's nothing here, so they nail the boards back on and then my dad says, well, it'll probably never happen again. Don't worry about it. If it does, let me know and we'll check it out again.

R: So everybody was happy and that was the end of that. I saw Jose during the week at the Post Office and I says, hey, you need to come and get that tesoro because there's too much happening. So he came over and got the tesoro, took it home and put it with one other stuff underneath his house. At that time we had space under the buildings, due to flooding. Jose put the tesoro in some boxes underneath his house and that's where it lay until 1963 when he went back after he had moved to California. He had moved to California in 1953 or 54.

R: In 1963 he went back to repair his windmill since he had purchased some used windmill parts. The caretaker drank a lot, and Jose had a hard time finding him. Jose decided to take all the boxes home to California and put them in the attic in his garage. Most of the contents of the boxes were old dishes, bottles, odd papers, letters, magazines and useless junk, and that's where it all remained until 2001-2002, when I met his son the internet, and his son informed me that his dad's name was Jose and was from San Antonio, and I called him and rekindled our youthful experiences and the discovery of an object shaped like an avocado, that had crashed on the ranch when we were little kids looking for a cow that was ready to have a calf. In one of our phone conversations, I asked Jose, what in the heck did we call that piece that we took off that object. Del Oro, Socorro, Ah, Tesoro. Yes, that's it. Tesoro. You know what, Jose says, I bet you it's still there, way back in the attic, its been so long that I had forgotten about it. Let me take a look and see if I can find it.

P: This the same piece that you remember, that very same piece? Okay.

R: Yes. Tesoro. He found it and so he FedEx'd. it. Right?

R: Yes. Because I wanted to get it examined.

END OF SIDE B, TAPE 1

Phone Conversation With Jose Padilla

P: Hello Jose, I heard that you'd love to do a dig with Reme? You want to do it soon because of the water that erodes the ground. There might be pieces there. Maybe people might b walking around over there.

J: born and raised there, and outsiders too.

P: Oh, they were born and raised there?

J: Some were, some weren't

P: Oh, no. I know you do. I know you know what happened, but do you think they're not tourists, right? They're just curiosity seekers?

J: Curiosity seekers and hunter, and there's a lot of rifle shooters ding target practice there.

P: Well, that's going to be fun.

J: That's just target practice; they set up targets, like bottles, and cans, I can understand city folks would have a hard time understanding this.

P: I know, but they're not target practicing on this stuff, are they?

J: No, no.

P: No, I know that.

J: Yes.

P: Have any of those guys approached you?

J: No, they were probably just target practicing, but it's a big area.

P: You're the one that actually took that crowbar or whatever you took and pulled that piece
   Off, right?

J: Right. I'm the one that used the cheater bar to jerk the piece off the panel.

P: And you saw the panel it was on.

J: Yes.

P: Yes, it's here somewhere.

J: You can see that, that it's clean, the pieces that are there now.

P: This would be 62 years later. Yes, you know it's a lot of years.

J: The piece was taken care of.

P: Great. Was there any of that funny metal that Reme had thrown into the crevice?

J: No, kept that.

P: Oh, okay.

J: It was on his "Tesoro". His treasure.

P: So is there any more of that anywhere?

J: I don't think so.

P: Well, we were little kids then. I was only about three feet tall. They were a little shorter than me-kind of light gray.

P: Do you remember if they looked directly at you?

J: I don't know whether they were looking directly, I know they were running back and forth, .P: Well, that's actually a good word because Reme was saying they seemed to be squealing in pain.

J: I think they were hurt.

P: Okay.

J: Because I wanted to go in there to help and

P: You were the 9-year-old, right?

K: Right.

P: But you had three days' worth of going back and forth there.

J: We had two days. Okay.

P: Oh, two days.

J: We had a whole week when they cleaned it up.

P: You had a whole week?

J: We used to take our horses on that ridge.

P: Did they see you? They didn't se you, did they?

 J: We were pretty sneaky, I know my territory.

 P: You were pretty sneaky. So you saw the soldiers throwing stud in that ditch, too?

J: Yes, whatever they didn't want to pick up, I guess, those pieces were too many.

P: Well, if you feel very good about going ahead with all of this, you, know, it's your story and Reme's , and my job is just to do an interview. I'll just do my best to record it word-for-word what you say, what Reme says and I'll have it transcribed. And thank you very much for being willing to do that.

J: I'm glad I can help.

P: I know. I very happily will do this and thank you for trusting me, and you can trust me because my work is all to just record and archive it for the future, so that's what I want to do.

J: There'll be no problem.

P: There'll be no problem. Oh, that's easy.

J: It's not the target shooters, the ground needs to be warm. I'm sure some pieces are still there after all this time.

P: Well, we will look into this part in the future. Right now, thank you for the testimony and your verifying the 1945 San Antonio Crash.

The Baca Sighting, Virginia and Reme Baca

P: Paola
R: Reme
W: Wife (Virginia-Jeannie)
P: And I can talk about this? When did this happen?
R: In July of 1994.
P: July of 1994, Okay.
W: It was a Sunday night.
R: It was a Sunday night.
P: It was a Sunday night and did you see it from here?
W: No, we were living in Tacoma. Across the Narrows Bridge.
P: So you were living in Tacoma and then what made you go outside?
W: It was hot. I was watching TV and he was out side sitting on the back porch steps. Finally he came and got me, said you have to see this, something's incoming. So I looked at the time, it was about 5 minutes to 11. So I went out there and of course then I saw this little ball of light coming towards us.
P: Can I take a picture of you talking to me?
W: Okay.
P: Just keep talking to me. I don't want to impose or anything.
W: So it was heading towards our house and it was coming down from the sky, heading in our direction. And
P: Okay, It was coming down from the sky and you got there just in time to see this little ball of light.
W: Well, it was still traveling down, He saw it from far away and then when I came out it was getting closer then and when it was almost at the house, you know, I was watching it. Couldn't figure out what in the heck it was because I never seen a ball of light travel by itself, you know.

139

P: So when you first saw it, it was like a….

W: A white orb. Yes, it was round but it didn't have any definite shape. It was like a beam of light, in the sense that it didn't have an edge on it. It just kind of feathered out and then when it got almost to our house, I'm watching it come over, it was starting to go over and all of a sudden a ship appeared. And I was telling…

P: Did it appear inside it or did it dissipate?

W: It appeared to the side of this orb.

P: Oh, to the side of the orb.

W: Yes, Yes, it was following the orb.

P: It just sort of appeared? The ship?

W: It was just there. All of a sudden it turned its lights on.

R: I'm looking at the other stuff up in the sky. And the reason I watched was because I had an insurance office right by McChord Field. And so I insured a large number of servicemen.

P: Okay.

R: And so I was following the air traffic back and forth..

W: A lot of airplanes went over where we lived during the day.

R: So that's what I was following when I saw that little light coming up like that. Why don't you watch for me, I asked Jeannie. I continued looking towards the base, and a moment later I asked Jeannie how she was doing and she said. "Look over your head".

W: Yeah, you did see it then.

R: You followed it.

W: I followed it, saw it all of a sudden disappear and then the ball of light's here and the ship is to the side so it maneuvered around more straight to go over our house, following that orb. The orb was always ahead of it a certain distance.

P: And so when it flashed on or when you were under it…

W: Cause I saw the light…

P: How long do you thing you were looking at it?

W: We watched it about 15 to 20 minutes.

P: Minutes? Fifteen to 20 minutes? That's an incredible, slow. Was it going slow? I mean, here you have this huge craft above you, moving just like you know.

R: It took that long to go from this edge to the other edge.

W: I could feel that they saw us. I could feel it, you know. And I could feel that they knew we were down there.

R: That long...

P: Twenty minutes...

R: That long being under, being covered.

W: And everything got real quiet all of a sudden when this thing...it was eerie. It was like you couldn't hear cars, no cars went down the street. It was, everything just blotto. Except for this.

P: Okay. So it took minutes to come really slowly...

W: Not completely 20 minutes. It was the whole thing, watching it from 20 minutes to going back...

P: When you did this and it was over your head, and you lifted your head and you saw the huge underbelly that was orange...how many seconds was it there?

W: Well, it was probably about...

P: I mean, did you see it and...

R: It was 20 minutes.

W: No, no

R: Not under...

W: Because we watched the whole thing go right back out.

R: But the edge comes in. When it comes in you . . . then very slowly it almost like it stopped.

W: No, no. Because I worried about that.

P: I know. Did you check the clock?

W: Yes when I went back in the house.

P: You did. And do you remember?

W: Yes. A 20 minute experience.

P: Do you remember what time you saw?

141

W: Five to 1v1 I went out and quarter after 11 I was back in the house.

P: You were back in the house and you actually could, like if you sat there you could see it for a
 While. You never were...

W: Oh, yes.

P: You could enjoy the show.

W & R: Oh, yes. It was a good show.

P: Okay. But then now, you know that's real. And you know you lived it. What about the Air Force base? Did they send anybody?

W: No, Because it's a funny thing. I' watching thing going back out again, having had enough, and of course the only way we could see the top of it was when it got out far enough that I could see the top of it. And as it was going out, it passed one of those big cargo airplanes.

P: C-47?

W: Yes.

P: It passed it?

W: Yes. It passed it. So the plane was going this way and he was headed for McChord and the thing just passed it and it didn't go fast.

P: That is really weird, because usually when there's some sort of sighting or some kind of anomaly, you have F-16s go right after it. Did you expect McChord would send some F-16s after it?

W: I didn't know.

R: I didn't, because I work with the pilots and my first instinct was when I went to work the next day....

P: You asked...

R: You know, they came in. These guys come in all the time. And one of the things that I did was, since they were members of NCOA and some of this group, I took care of their insurance needs....

P: And what's NCOA?

R Non-Commissioned Officers Association.

P: Okay.

R: When some of them came in, I told them what I saw and asked if they were familiar with any of the UFO activity?

P: You actually said that?

R: Oh, yes, and they said, "we call them bogeys.

P: Okay.

R: And I said that I needed to report one. And the response was "We don't take any reports". Anything that goes over 10,000 miles per hour we don't touch. He says we don't do it, but call such and such. So then I called that number at the base and they said, we don't take them, but here's some phone numbers that you can call and they will...and then when I called.

R: No, no But going back, in fact I think you've mention him, that you had lived in the same city or knew his wife there in Arizona...

P: It wasn't McDonald...

P: Allen Hynek? You called Allen Hynek? That's the man I worked with for six years. You called Allen.

R: Yes. I called Allen Hynek.

P: Oh my God.

R: And so I talked to his son.

P: He wasn't in New York though. He was in Evanston, Illinois. Yes, it was Evanston . . . The Center for UFO Studies.

R: Yes,

P: Oh my goodness.

R: Remember, I'm a green guy... I don't know stuff about UFOs at that time.

P: No, no, no. I know. Can you tell me the date again of this?

R: How about like July 7, makes a lot of sense.

P: Of what year?

R: 1994

P: He was dead by then. Yes, he was dead.

R: His son told me, my dad passed...

P: He died in 1986. But the Center was still going with Mark Rodeghier.

R: So he took the information and basically that was it.

P: Okay. Do you have that telephone number still? Because I don't know if it was Paul that you talked to, Paul Hynek. I know very well, I mean he had five children, but I don't know which One...

R: Anyway, I didn't know. I was not into this UFO stuff.

P: You just did it because you felt the need to do it...

R: I had been given this number by the Air force, and was following up, All of a sudden in a few minutes of this night, some gigantic changes take place in our lives, and now my wife is more understanding with regards to ET.

W: Also, I was going to wake up the neighbors but we had a fence...

P: Did you have neighbors out there?

W: They had just gone to bed.

P: Oh, because of the time.

W: Yes, because of the time and if it wasn't for the fence I would have gone and knocked on their

R: She worked for the fire Department.

W: But it meant that I would have had to go to the back door. I was barefooted, so it meant that I would have to go the back way on gravel in my bare feet and trample around and you're doing this at night in the dark, and I was afraid I was going to hurt my feet, so I couldn't get across there.

R: And they were doing some remodeling of the house so they had boards with nails and roofing on the ground.

P: So when you first heard about Reme's story when he was very young, did you just feel that it was against religion to talk about it?

W: No. I just didn't believe. Period.

W: No, I didn't believe in UFOs.

P: Then you thought he might have found an airplane or something?

W: You know, he never went into it.

P: Okay. So now that you're here, because he does go into it in great detail now, so he just didn't talk to you about it in great detail.

W: No.

P: And you didn't talk to her about it because you thought you'd scare her?

R: No.

W: No, I didn't believe it. I mean, he talked to me about a UFO like that time, I said...

P: But he never lied. He doesn't lie.

W: I know that, but I...

P: Your mind couldn't wrap around it.

W: I couldn't fathom anything like that.

P: Your mind couldn't go there.

W: And then when you see it, oh God...

P: So when you saw it and did you feel like you had a transformation when you saw it?

W: Well, it changed my thoughts in what's out there. Completely.

P: Did you get any feelings when it was over your head? What did you get?

W: Calmness, peace.

W: I wasn't scared.

P: You weren't scared?

W: No, not at any time. I was leery that it was going to stop...because I've heard, you know, you see stories about abductions and things like that, and so I was...as long as that thing was moving, I was all right. But if they had stopped, then I would have been long gone.

P: Okay.

W: Because I have never seen a ship that size go that slow. Our airplanes don't. that was slow for the size of that craft.

P: You know, I really recommend you see Close Encounters of the third kind, the movie, because that ship...

W: I did.

P: Did you? That ship comes in total silence, and just...

W: Yes I know. I saw the movie. We got the video, still have it.

P: So you were watching Close Encounters happen underneath it. Looking underneath it.

W: Oh, yes.

P: You got a feeling that it was calm. What else?

W: Right, right.

P: You told me before you think that they were watching you.

W: Yes, yes, I had...because of where they were and where we were. Why are they going exactly overhead of us, you know. It's like they specifically came to visit us. This seemed to be done on purpose, in a sense. Because there'd be no other reason why they'd come. I was fascinated with the colors. I've never seen anything so bright in my life as far as color-wise.

P: What was underneath? Were the colors moving or were they more like...

W: No, just a solid glow. It had marbling of this dark in it, like marbles around it. But mostly it was really a deep red-orange. Not real red and not real orange, but what you would call a red-orange.

.P: Okay. And other feelings you got?

.W: Well, as it kept going I felt more comfortable that it wasn't going to stop after it went over our house and we watched it go off, head back up again. So then I went in the house and cried.

P: You said you went in the house and cried.

146

W: Yes.

P: It was just an emotional experience.

W: It was just emotional, yes.

P: Then (Reme) what did you feel when you saw it.?

R: It reminded me a lot of the San Antonio crash, the eerie feeling, because when Jose and I were there at the crash, it seemed like the world stood still, birds quit chirping and the dogs quit and all those normal things appeared to be in suspension.

W: Yes, nothing moved. No cars, no nothing.

R: And you've seen where we lived, we lived across the street from the Catholic Church. This is a business district, there were car lots about a block away, restaurants, cafes, and other stores. The fire department is about two bocks away from our house. The police patrol the area. There are normally police cars driving back and forth thru the business district. It is well lighted up so nobody can hide in the shadows, and there is always a lot of traffic out there, a lot of people walking back and forth. And that particular night, if anyone called out for help from the authorities, it might not have been forthcoming.

P: They stopped time and space for you, because those cops were still running around and everything was still going but you didn't feel it. What did you feel, though, when you saw it? Did you just say, "Oh, I know this is real because I already had this?" What did you feel? Or were you worried about her?

R: No. Well, it was something new for me actually. It was new for me, too, because I have seen a craft like this , before either. One thought did g thru my mind. Did those creatures survive the San Antonio crash, and now they are grown up, in the area and decided to say high. But on the other hand, if I follow that line of thinking, How did they know where I lived?

P: Well that's a possibility.

147

R: Yes. But it was at night. I had never seen anything like that.

W: But I got to see the top of it and I saw what was on top. So, there was a mound on top.

P: A mount on top. Could you have seen a picture of a UFO...

W: Yes, it wasn't completely flat, like a complete disk. It just went...

P: Do you think from what she drew that it looked like the one that that crashed?

R: No, completely different ship. And ours wasn't that big. It was 25 to 30 feet.

P: This one you would say...

R: This one wasn't all that far. It was about 200 to 300 feet up above us.

W: It covered the whole sky.

R: It covered the whole sky.

P: It covered the whole sky.

W: I could see no blue or dark. It was night time.

P: So the circumference of it was....

W: Oh, God, it was like a football field.

P: The size of a football field.

R: Very big, it would crush the church.

W: Well, when it went beside that plane, you knew it was big. Oh, yes, because I saw the difference in the size of that plane and the size of that craft.

P: How many times bigger. Would the craft be than the plane?

W: That was a big cargo plane. How big are those?

P: How many of those planes would make that craft?

W: It looked similar in size to me, because now you're getting high up in the sky.

P: So would you say the UFO was five times the cargo plane?

148

W. The comparison I was doing was about seven miles away above McChord Air Force Base after the craft left our house and they passed each other. The best comparison would be when the craft was about 200-300 feet above our house. It was as big as a football field. If it had had landed, it would have covered the church, our house and our neighbor hood. That was a pretty large craft. I am aware of the Size of a cargo plane. There is no comparison.

W: At the distance of about seven miles away, they looked equal as they passed one another.

P: Oh, they did?

W: Yes.

R: You're looking at the horizon.

P: But one is the plane and one is…it's amazing they didn't know each other was there, I mean…

R: Oh, they knew. I bet they saw…

P: You think they did see it?

R: Yes they did see it.

W: How could you miss something like that?

P: Well, because it's another dimension and I don't know how they could miss it but nobody's talking about it, right?

W: No.

P: They never do. This is a bogey and they're not allowed to do it. Is there anything else about the experience that you remember? You were happy because then she changed her way of thinking and this would allow you to come out with your story.

R: Well, to begin to research it. Find out….

P: Okay, so 1994, then ten years, more than ten years….

W: He had to find Jose. After he found Jose, then everything started coming out.

P: When did you find Jose?

R: You know, what my problem was, was trying to recollect from way back then.

P: But what year do you think you found Jose?

149

R: In 2002. It was after my surgery. I had open-heart surgery. It is obvious that finding his childhood friend Jose changed Reme's life and his current reality, carrying the burden of truth takes its toll on most families. It exposes them to fear and ridicule. It was so fortunate that Virginia, through her own sighting, has become the steadfast supporter I often wonder if that event was intentional, if the ETs are encouraging some type of disclosure by entering into this human dimension. It is a lonely existence to live in this alternate reality without the support of family and friends.

Interview with William Brophy Jr.

PH: You're still in Florida and how did Italy go for you?

WB: That was nice, they invited me to talk about the 1933 crash.

PH: You mean the Italian 1933 crash. How did you find out about that?

WB: There was this story Dad told me about this bomb that he had witnessed, the Trinity bomb. You know the crash in Italy on June 13, 1933. That's San Antonio day there.

PH: 1933? I know about Mussolinis' X-Files and I know Pinotti very well, I used to work for him. What does that have to do with Trinity?

WB: The crash in Italy involved a scout ship like Adamski, you know? With the small alien with oriental features.

PH: And did your dad tell you that?

WB: Yes. And the Japanese attacked Pearl Harbor on December 7, and you know they had beings in their legends. They believe they're descended from these people.

PH: What did the Japanese call these people?

WB: Nordic aliens, with blue eyes.

PH: What does Mussolini's 1933 crash have to do with the Trinity site? Did you find out anything from that?

WB: Enrico Fermi, some of the information he used for his chair reaction experiment in Chicago came from Italy.

PH: Fermi was down at the Trinity site.

WB: Yes, at Alamogordo. My dad was there, on the flight line to Alamogordo when that happened.

PH: So your dad was very much involved. Did he know about the 1933 crash in Italy?

WB: Yes, because he was interviewed by Eric Henry Wang, who was in charge of the aliens when they brought them back from Italy.

151

PH: Where did they bring those tall aliens?

WB: Into the Wright Patterson Air Force Base.

PH: So in 1933 they brought back tall, blond aliens from the crash in Italy.

WB: No, in April 1945, when we captured the Savoya Marchese Avation facility in Italy, that's when we brought the UFO and bodies back.

PH: So those were bodies that went back to Wright Patterson.

WB: Yes, and a scout ship too.

PH: Were any photographs ever taken of them?

WB: Len Stringfield got it all in the retrieval book in 1982.

PH: That's not Situation Red, is it? What's it called?

WB: I don't remember, it's Len Stringfield's retrievals book from 1982.

PH: And he talks about it there.

WB: Yes, these two very tall blond aliens with oriental features on their eyes.

PH: You know so much about this, did your dad just sit you down and talk to you about this?

WB: Well, he was involved in San Antonio New Mexico, the UFO retrieval there.

PH: But did he sit down with you and tell you all about this, because I know they were sworn to secrecy.

WB: We were watching a program in 1978, he was not longer in the service at that time, regarding Roswell. Remember In Search Of? My dad thought they were declassifying stuff, and so I asked my dad whether this was true about these UFO cases, and he told me about it.

PH: So he thought they were declassifying it.

WB: Yes, because it was there on TV. They had Jesse Marcel there talking about it, and he thought they were declassifying the whole thing.

PH: Then, what he did was, you said that your dad was stationed near San Antonio at the time, at Alamogordo. What was he doing there?

WB: On June 1942 Alamogordo Army Air field (AAF) was established at a site six miles (10K) West of Alamogordo, New Mexico. Initial plans called for the base to serve as the center for the British Overseas Training Program. The British hoped to be able to train their aircrews over the open New Mexico skies. However, everything changed when the Japanese launched a surprise attack against the Hawaiian Islands on 7 December 1941. The British decided to no longer pursue its overseas training program, and the United States military saw the location as an opportunity to train it own growing military. Construction began at the airfield on 5 February 1942 and forces began to move into Alamogordo Bombing and Gunnery Range on May 1942. The base was under the command of the United States Army Air force(USAAF) Second Air Force with its headquarters at, Colorado Springs, Colorado. The base was runways, taxiways and hangars during the summer of 1942 being renamed Alamogordo AAF in June. From 1942-1945, Alamogordo AAF served as the training grounds for over 20 different groups, flying initially Boeing B17 Flying Fortresses then Consolidated B-24 Liberators. Typically, these groups served at the airfield for several months, training their personnel before heading to combat overseas. The host support unit at Alamogordo AAF was the 359th Base HQ and Air Base Squadron, Activated on 10 June 1942. This was re-designated the 231st AAF Base Unit on 25 March 1944, then 1073d AAF Base Unit on 24 August 1944. On 16 April 1945 Alamogordo AAF Was relieved of its training mission and assigned to Continental Air Forces, and was scheduled to be a permanent B-29 base. However postwar funding cutbacks did not allow an active bomb group to be based at the facility, and the base was temporarily inactivated on 28 February 1946.

WB: He was flying B-29 Bombers, you know the bomber we used over Japan?

PH: He was just flying them. Did he get involved in any UFO crashes while he was there?

WB: This was also a training facility, training pilots and their crews on the latest instrument additions or changes. During this time there was a new lens instrument that had been added for the bombing efficiency, and so the pilots and their crews had to be trained on it. Yes, he was involved in the August 1945 crash in San Antonio.

PH: Was he involved in any others?

WB: He had an encounter with a big cigar in 1950.

PH: He had an encounter with a big cigar. So when he was stationed in San Antonio, where were his quarters? Were they at Alamogordo?

WB: He was not stationed in San Antonio, there are no military bases in San Antonio New Mexico. He was stationed at Alamogordo Army Air Force Base.

WB: He was a member of the 231st.

PH: So he was stationed at the Alamogordo Army Air Force Base when this crash happened, that Remy and Jose saw.

WB: August 15, 1945, just after Trinity.

PH: What did he do, did they call the army to clean it up? I imagine it was the Army because there was no Air Corps at that time. Did he go in there, or what did he do?

WB: A B-49 bomber crew on it last training mission going over Walnut Creek had reported smoke in that vicinity. There was a communications tower that had been destroyed by something. They thought a plane might have hit the tower and crashed on this ranch out there. My dad's commanding officer, Colonel Maurice Preston was briefed on this. They were under the impression that it might have been a plane, but when they finally got out there, they found something a little different. The thing had a kind of a fin on it. They thought it was dome kind of an aerial vehicle.

PH: It had a fin on it?

WB: Yes, it was kind of an oval shape with a fin on the back end.

PH: Ok, because they didn't describe the fin but they did see the oval shape. So your dad went out there because of the tower. Is there any way that I can find anything relating to Maurice Preston anywhere?

WB: No, he didn't go there because of the tower. He went there because of duty. General Maurice Arthur Preston commanded the 41st Bombardment Wing from October 1944 until May 1945, when he returned to the United States. General Preston was then assigned as base commander of the 231st Army Air Force Base Unit at Alamogordo, New Mexico. He enrolled as a student at the Air Command and Staff School, Maxwell Field, Alabama, in August 1946, and upon graduating in June 1947, became the Chief, Inter-American Security Branch and Military Coordinating Committee, of the permanent Joint Board of Defense Canada and United States, Washington, D.C. He later joined the Plans Division of Air Force Headquarters in Washington, D.C. He became commander, U.S. Forces Japan and commander, Fifth Air Force, in August 1963. These are just part of General Preston's accomplishments. The Air Corps found the Wreckage, the Army under the command of Colonel Harold R. July 9, 1945-August 3, 1947. Turner did what we today refer to as a recovery operation. 1st Commander White Sands Proving Ground, Col Harold R. Turner, (then Lieutenant Colonel) was the first Commander of White Sands Proving Ground. He is quoted as having said, "I had arrived in Las Cruces, NM, two days before the explosion (16 July 45). When the bomb went off, I was asleep in the Amador Hotel in Las Cruces. The first I knew of it was on reading the morning paper, although it was on the property I was supposed to command." Turner had the task of getting the proving ground ready to test rockets. There were no buildings, potable water or transport facilities in the area. Not only was there nothing at the site, it was still called "that place in the desert." Turner is credited with naming it White Sands Proving Ground so personnel would have a mailing address. Through his efforts water was found and a reservoir and pumping system built. Working with the Corps of Engineers in Albuquerque, housing, dining, and sanitary facilities were built. It was Turner's decision to put the Administration area at the foot of the Organ Mountains, two miles from the first rocket firing station. After the war ended Turner was informed 300 railroad carloads of captured German V-2 components were in Law Cruces. He was told to get the Proving Ground ready to fire them. Though he started with no transport facilities and little money, he moved the components by truck to the proving grounds over the San Andres Mountains and developed a shop to put them together. The first 28 V-2s were fired during his tenure. After Turner left the proving ground, and soon after his promotion to colonel, he was awarded the Legion of Merit. The award was presented for his work both in the New Mexico desert and the almost equally difficult one of preparing a precision range for aerial rocket firing on marsh land near Dover, Delaware, from June 1944 to June 1945. The award citation read in part: "Lieutenant Colonel Turner performed exceptionally meritorious services...as commanding officer, White Sands Proving Ground, Las Cruces, New Mexico, from June 1945 to July 1947. Colonel Turner assumed responsibility for the planning, building and operation of the White Sands Proving Ground for controlled missiles, and under his

156

personal direction this proving ground was built in a remarkable short time in a remote and uninhabited desert area and was completed in time for the first scheduled test firing of important rocket missiles. In 1974 the Nuclear Weapons Effects Laboratory was named for Turner. During the dedication Ceremony, Maj. Gen. Arthur H. Sweeney, Jr., Range Commander, said, "Colonel Turner was an educator, innovator and above all leader. We must remember there were no precedents or policies for him to follow. This great pioneer did what no one had ever done before he established an installation that would be the birthplace of America's missile and space activity.

PH: So can I go on the Air Force Web site and find him?

WB: Yes, he's a real famous Air Force general. They've got lots of information on him there.

PH: Did your dad talk about him being a friend or anything?

WB: Yes General Preston was my dad's Commanding Officer.

PH: So when your dad told you about this, you were watching TV?

WB: I wasn't watching TV every time we talked about it. It was the TV program that originally initiated the conversation..

PH: What did he say, that he was called there? What did he see?

WB: The M.P. said he had this little being in custody, and it looked like a little man but it also looked like some kind of a bug. I had these praying mantis type features on it, real strange eyes. In other words, it looked like a small man with a big head but these bug like features. It was real weird looking.

PH: The bug like features, was that mostly the eyes?

WB: Yes. The eyes, the face. It was kind of like a little grey, but with bug like features.

157

PH: That supports Remy's story, who said it was like a mantis type. How tall did you dad say it was?

WB: About four feet tall.

PH: Was it grey with long spindly arms?

WB: Kind of grey.

PH: Was it alive?

WB: There was one alive and two dead.

PH: Where did they go? What did they do with them?

WB: They took them to the base and then flew them off to Wright Patterson in a B-54.

PH: How soon did they do that?

WB: On the 16th.

PH: So the creature only stayed there until the next day. What did your dad say?

WB: They brought the thing back to the base and then shipped it out by truck. He said there was a lot of wreckage around the site that had a lot of what looked like Sanskrit or some Egyptian hieroglyphics, some kind of symbols on them. On a great big hole in the side of this thing, it had a lot of what looked like spider webs hanging down, but it was real strong.

PH: Spider Webs, which is what Remy describes as angel hair, right?

WB: Yes, it's real strong, real light, but it will wave in the wind.

PH: Was it white?

WB: It looked like a spider web.

PH: When Remy and Jose tell the story, they say that when Jose jumped inside the ship and pulled the panel out, it had all been cleaned out.

WB: Soldiers were picking up all the loose stuff all around.

PH: Yes, but on the inside?

WB: My dad didn't go inside it, he walked around and picked up all the pieces that were lying around.

PH: So your dad had to clean it up?

WB: No, he was an Officer, the enlisted men were picking things up, he was just helping. I don't believe there was any protocol for this type of activity in 1945.

PH: But your dad was a pilot, and he was in charge of the clean up?

WB: As the highest ranking officer in the field, he was temporarily in charge by the direction of Colonel Preston, until the following day when Army Col. Turner took over.

PH: Did your dad see anyone else, who wasn't part of the cleanup crew?

WB: No, just the military people there.

PH: I heard that he had seen two kittle Indian boys. Did he actually say that?

WB: That was later on.

WB: By the Owl Bar, in San Antonio. That's this bar where everyone would go.

PH: What did you dad tell you about the Owl Bar?

WB: That's a famous bar where everybody would meet, like a restaurant. Oppenheimer would eat there.

PH: Do you think Oppenheimer saw the being?

WB: No, because he left after Trinity.

PH: It was only a month after Trinity that this happened. Maybe he could have been there?

WB: Enrico Fermi used to go there, a lot of scientists.

PH: Did your dad go there too?

WB: Yes, my dad and his friends went there all the time. They used to go to Socorro, to the USO. (United Service Organization)

PH: Now about these Indian boys, where did that come from?

WB: Oh, that was at the Owl Bar, I think…

PH: You had told Remy that your father had said something about seeing two little Indian boys.

WB: Yes, that was during the cleanup.

159

PH: Did your father say anything else that you an remember about the cleanup or about the crash?
WB: Just that there were a lot of pieces, and it was funny looked like Sanskrit or like Egyptian hieroglyphics.   It's real strong stuff, too.
PH: How old were you when your dad told you all this.
WB: This all begin in 1978, so I was about 23.

End of William Brophy Interview regarding San Antonio Crash.

Chapter Nine

Lt. Col. William Brophy

Pre Bomb Crash Site

Jumbo

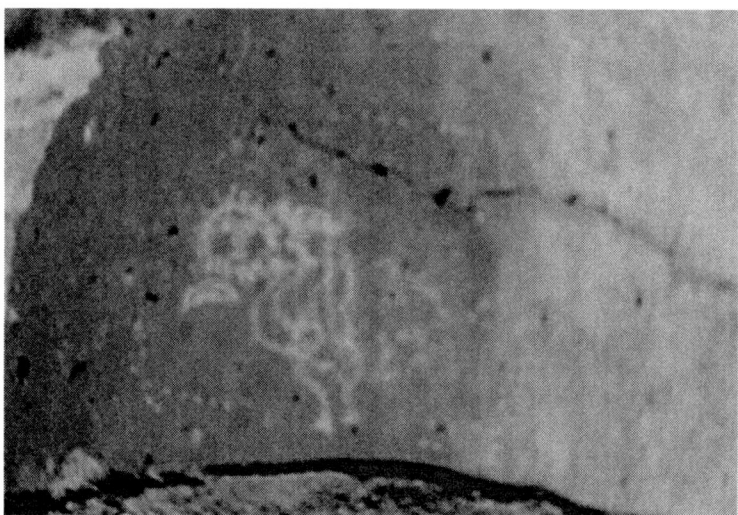

Pictograph

Entrance Gate to Crash Site

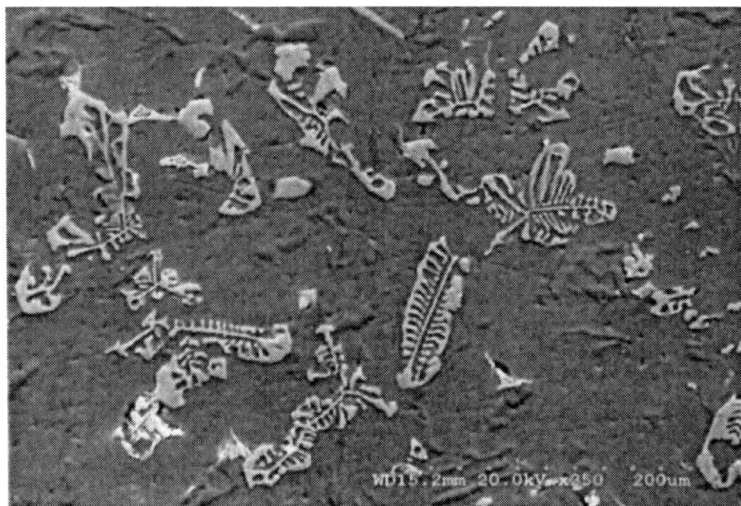

The photo of the casting has a scale at the bottom. It is a total of 200um or 200microns. Between each dot is 20microns. The average human hair is 40-120microns, so at 60microns or 3 spaces on the scale you can see that the fish skeleton looking thing in the middle is narrower than a human hair. It also means that the straight lines are probably less than a fortieth the width of a human hair since they measure about 2microns in width. I am willing to bet that a casting detail this fine could not be done in the 1940's. These casting lines are very significant because they would be "outees" on the mold in order for them to be "inees" on the casting. Today it can probably be done, but not back then.

Casting Lines

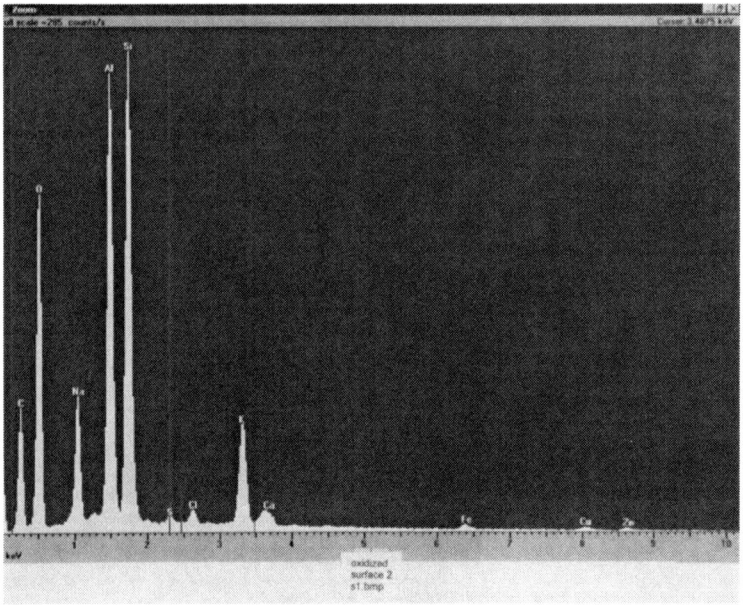

Metal Analysis

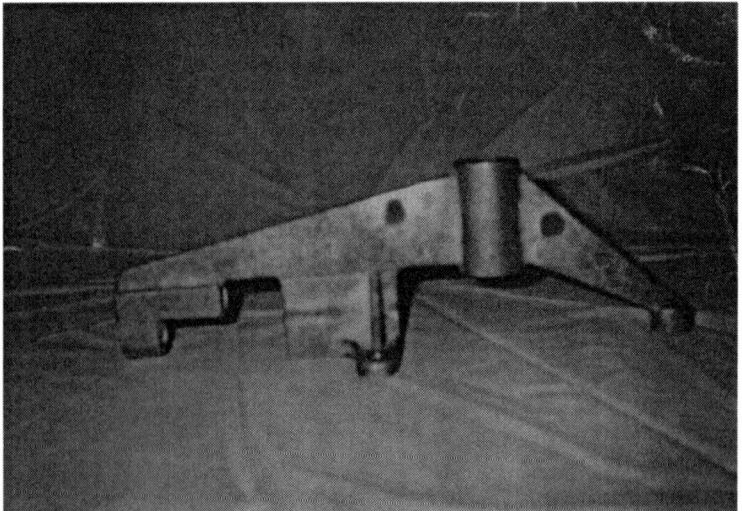

Bracket

166

Owl Bar and Café

Jose and Reme

167

Reme's Residence

1909

Hilton Store

BIBLIOGRAPHY

http://hanford-downwinders.tribe.net/thread/788190be-1d41-4e64-

13-Committee on an assessment of CDC Radiation Studies, National Research Council, A Review of Two Hanford Environmental Dose Reconstruction Project (HEDR) Dosimetry Reports: Columbia River Pathway and Atmospheric Pathway (National Academy Press, 1995), p.2.

LOS ALAMOS beginning of an era 1943-1945
Information for this story was compiled and edited from previously published articles and brochures on Los Alamos history written by the staff of LASL's Public Relations Office

Is It True What They Say About Dixy? A Biography of Dixy Lee Ray by Louis R. Guzzo Published by the writing works, Inc 7438 S.E. 40th Street Mercer Island, Washington 98040

The Day After Roswell Col. Philip J. Corso, (Ret) With William J. Birnes Copyright © 1997 by Rosewood Woods Productions, Inc. Forward copyright © 1997 Senator Strom Thurmond.

Global Fallout IDEALIST.WS

Trinity Atomic Website by the U.S. Department of Energy National Atomic Museum, Albuquerque, New Mexico
Kay Sutherland a cultural anthropologist with S.T. Edwards University Austin Texas
Exopolitics Stargate to a New Reality © 2011 Paola Leopizazi Harris. All rights reserved.

CPSIA information can be obtained at www.ICGtesting.com
Printed in the USA
BVOW030123041012

301975BV00007B/8/P